FISSION PRODUCT YIELDS
AND THEIR MASS DISTRIBUTION

VYKHODY PRODUKTOV DELENIYA
I IKH RASPREDELENIE PO MASSAM

ВЫХОДЫ ПРОДУКТОВ ДЕЛЕНИЯ
И ИХ РАСПРЕДЕЛЕНИЕ ПО МАССАМ

FISSION PRODUCT YIELDS AND THEIR MASS DISTRIBUTION

By

Yu. A. Zysin, A. A. Lbov
and L. I. Sel'chenkov

Authorized Translation from the Russian

CONSULTANTS BUREAU
NEW YORK
1964

The original Russian text was published for the State Committee on the Uses of Atomic Energy in the USSR by Gosatomizdat (State Scientific and Technical Press for Atomic Energy) in Moscow, in 1963.

Ю. А. Зысин, А. А. Лбов,
Л. И. Сельченков
Выходы продуктов деления
и их распределение
по массам

Library of Congress Catalog Card Number 65-10525

ISBN-13: 978-1-4684-0666-5   e-ISBN-13: 978-1-4684-0664-1
DOI: 10.1007/978-1-4684-0664-1

CONTENTS

INTRODUCTION

Since the time of the discovery of the fission of the nucleus, when Hahn and Strassmann first detected the products of nuclear fission in 1939 in uranium irradiated by neutrons, over two decades have passed. In the work accomplished since that time, research has been greatly expanded on fission products and has been extended to many nuclides formed in the fission of various nuclei induced by neutrons, gamma rays, charged particles, and in spontaneous fission.

Amassing of experimental results is an important task not only for nuclear engineering but also for an understanding of the fission process itself, for bringing into existence a finished theory on the unique and intriguing physical phenomenon which plays such a decisive role in the practical utilization of nuclear energy.

In this reference manual, an attempt is made to collect and generalize upon the results of experimental research work over the period spanning 1939 to 1962, on yields of fission products and on the mass distribution of fission fragments. Various instances of the fission of nuclei induced by neutrons, gamma rays, charged particles, and spontaneous fission of nuclei are discussed.

In review papers published on the yields of fission products [1-6], only certain distinct and special topics are taken up. These reviews have been based on data which are not obsolete. For example, the most complete review, of those published, on fission fragment yields in neutron-induced fission [6] (a review appearing back in 1960) did not include data on fragment yields in the fission of Th^{232} by 14 MeV neutrons, fission of U^{235} and U^{238} by 8 MeV neutrons, fission of Am^{241} by pile neutrons, etc. The fragment yield data cited in this review for cases of fission of U^{235}, Pu^{239}, U^{233}, U^{238}, Th^{232} by thermal neutrons, fission-spectrum neutrons, and 14 MeV neutrons require additions and supplementation. Moreover, the review [6] quite naturally lacks the data published in the most recent years.

A multiplicity of experimental results on the yields of fission products of a large number of different fissionable isotopes in the most varied instances of fission of those isotopes by neutrons, charged particles, gamma rays, and in spontaneous fission, continue to the present date to be found scattered in any number of distinct papers published in different journals and different times.

The authors of the present work took upon themselves the task of filling this gap, in compiling the most complete summary possible under the circumstances of all the available experimental data on yields of fission products. The book collects between one set of covers all the data referable to fission events at excitation energies up to 100 MeV. The reason for this is that the specific features of the fission process at very high excitation energies are masked, to an appreciable extent, by additional effects (such as fragmentation) which go beyond the scope of this book. Furthermore, the reader will find a review [7] appearing in 1960 which contains data on the yields of products of fission by high-energy particles.

ABSOLUTE AND RELATIVE YIELDS
OF FISSION FRAGMENTS

In the tables presented below, absolute yields of fission fragments will be found listed. The term absolute yields* of the fragments as used here refers to the probabilities of formation of fragments of a specific type, expressed percentagewise, calculated per single fission event:

$$Y(\%) = \frac{N}{P} \cdot 100, \qquad (1)$$

where Y is the absolute yield of the fragment; N is the number of atoms of a given fragment formed in a specimen as a result of P fission events.

Fission fragment yields may be subdivided under three headings:

a) independent;

b) cumulative;

c) total.

The term independent yield is understood to mean the probability that a given isotope will form directly in the fission process.

The term cumulative yield is understood to mean the probability that a given isotope will form both directly in the fission process and by way of decay of the parent nuclide (which might have formed in turn from its own parent nuclide, and so forth).

The term total yield is understood to mean the probability that the final product of the radioactive decay chain for a specific mass number A will form. The sum of the total yields is 200% in binary fission.

Cumulative yields of the last links in the chain of radioactive decay are equal in practice to the total yields of the given mass number A.

---

* In some contributions to the literature, absolute yields are presented as cross sections of fragment formation (viz., products of the fission cross sections by absolute yields).

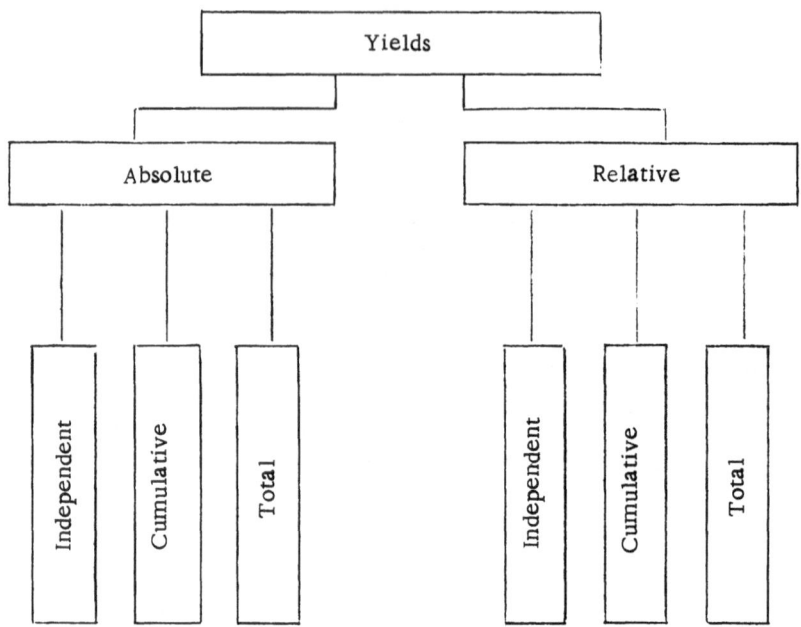

Fig. 1. Scheme for classifying fragment yields.

The relative yields γ_i must be distinguished from the absolute yields, the former term referring to the ratio of the absolute yields of different nuclides Y_i formed in fission to the yield of a selected nuclei Y (whose yield is assigned the value unity):

$$\gamma_i = \frac{Y_i}{Y} \; .\tag{2}$$

The classification of the fragment yields is illustrated in the block diagram above (Fig. 1).

METHODS FOR DETERMINING
FISSION FRAGMENT YIELDS

The first stage in all of these methods is to irradiate the selected fissionable isotope with neutrons, gamma rays, or charged particles of a specified energy. The irradiation time is selected on the basis of the sensitivity of the method used and the yield of the fragment of interest. The study of fragment yields resulting from spontaneous fission is of course an exception.

Clearly, from formula (1), it is required in all cases to find out the number of fissions in the specimen and the number of atoms of the fragment of interest present in the specimen. The total number of fission events is usually found by directly counting the number of fission events with the aid of a fission chamber with an exactly known amount of that same fissionable material in the thin layer of the chamber. The chamber is then placed in the same neutron flux in which the specimen irradiated has been placed. A standardized BF_3 counter is utilized to measure the fragment yields in fission by thermal neutrons.

Radiochemical techniques have been exploited in most of the work on fission fragment yields. These techniques require that the specimen of fissionable material be dissolved and that a quantitative chemical separation be made of the elements involved.

After the radiochemical treatment, an absolute measurement is performed of the activities of the corresponding isotopes, i.e., the fission products. The measurements are usually conducted using 4π-counters, even though end-window counters and even cylindrical counters had been employed in earlier work. Some workers did not determine the absolute number of fission events directly, but computed the absolute yields in terms of a reference isotope whose yield was assumed known or assumed equal to the yield known for some other case of fission. In some papers, the number of fission events was not determined, rather only the relative yields were measured. The error in the determination of the absolute yields by radiochemical methods, as a rule, to approximately 7 to 15%. The error in relative yields determinations was lower.

In some cases, particularly in measurements involving higher precision (to inquire into the fine structure of the mass distribution, or to study the burnup of individual isotopes, etc.), mass spectrometric techniques were resorted to. This approach, in contrast to the radiochemical approach, is also applicable in the study of yields of stable and long-lived fission products. Sometimes recourse was had to a preliminary chemical purification of the specimen with subsequent separation and isolation of the fractions of interest, in work with mass spectrometric techniques [8, 9].

The integrated mass spectrometric method [10], consisting in a preliminary determination of the relative ionization coefficients for the elements under study followed by measurement of the integrated ion mass currents, has also been used. Absolute measurements were carried out on the basis of relative measurements made with respect to one of the masses involved which was determined absolutely by the isotopic dilution method [11]. Some data were obtained by the "external standard" technique [12, 13], where the ion current integral for a given mass of a mixture of all fragments is compared to the ion current integral of the same mass of solutions of known concentration. In most cases, the mass spectrometric measurements displayed a relative character.

The mass spectrometric procedure has been exploited in many research efforts [4, 14-20]. The accuracy attained in determining the relative yields was 3-5% as a rule, using mass spectrometry.*

A third method for determining absolute yields has been described and utilized by V. A. Vlasov, et al. [21], and is based on equating 200% of the sum of all the relative yields arrived at by extrapolation from experimentally derived (by any method whatever) data points and "reflected" points. The accuracy of this method is further improved to the extent that relative yields have been determined for larger and larger numbers of the fragments involved.

Finally, in some of the papers the mass distribution of fission fragments was found from experimental results on the energy distribution of the fragments, obtained either by means of a double ionization chamber [22, 23] simultaneously recording the energies associated with both fragments, or by utilizing time of flight methods [24, 140].

---

* A paper by E. P. Steinberg and L. E. Glendenin [4] estimated a 10-20% error in the determination of absolute yields by radiochemical and mass spectrometric techniques.

The high accuracy achieved in mass spectrometric determinations* of fission fragment yields may be exploited in studying the fine structure in the mass distribution of the fragments and in studying the cross sections for neutron capture by fission products [4, 14, 20, 141].

---

* The use of mass spectrometric techniques in high-level irradiation by thermal neutrons has made it possible to compute the burnup cross sections of radioactive nuclides formed in the fission process [141]. These cross sections turn out to be quite appreciable [4, 14, 20]. For example, in the case of Cs^{133}, Xe^{135}, Nd^{143}, Pm^{147}, Sm^{149}, Sm^{151}, the burnup cross sections are, respectively: 40; $3.4 \cdot 10^6$; 334; 200; 84,000; 12,800 barns [20]; in the case of Xe^{133} we have 3000 barns [4]. The absorption cross sections cited by M. P. Anikina, et al. [14] for Pm^{147}, Sm^{149}, Sm^{151} are, respectively: 90 ± 20; $49,000 \pm 8000$; 8000 ± 1500 barns.

FUNDAMENTAL REGULARITIES
AND TYPICAL FEATURES OF THE MASS
DISTRIBUTION OF FISSION FRAGMENTS

The mass distribution curves of fission products, which can be plotted on the basis of the corresponding yields, as mentioned earlier, are of utmost importance for an understanding of the physics of the fission process.

Such distributions must be constructed, generally speaking, on the basis of the total yields for a given mass. However, bearing in mind the fact that the difference between cumulative yields of the last links of the chain from the total yields is substantially less than the corresponding errors (particularly when recourse is had to the radiochemical method), we find that the cumulative yields may be utilized.

In individual cases, corrections have to be introduced for the decay schemes (cf. the schemes of decay chains). For example, the yields of Cd^{115} and Cd^{115m} must be totaled in determining the yield of the mass 115.

In the overwhelming majority of cases, as we know, fission ends up in two fragments. Experimental results reveal that the corresponding probability of ternary fission (with the formation of α-particles) is roughly 10^3 times below that of binary fission [25]. The frequency of ternary fission is estimated in one paper [4] to be 1 to 3 per 1000 fission events.

Radiochemical searches for Be^7 and Li^9 have demonstrated that the yields of Be^7 and Li^9 in uranium fission are respectively less than $10^{-5}\%$ and approximately 0.05%, while the yield of Be^{10} is placed under serious doubt and is estimated at below $4 \cdot 10^{-4}\%$. The yield of Be^7 in fission of U^{233} by thermal neutrons is less than $3 \cdot 10^{-7}\%$, that of Mg^{28} less than $4.2 \cdot 10^{-9}\%$ [26]. This is evidently due to the extremely low probability of the critical state of the nucleus, required for fission into other than two fragments, to be achieved. On the basis of the above discussion, then, we may assume that the mass distribution curve of the fragments corresponds with complete accuracy to the distribution curve for binary fission.

Consider the fundamental regularities and the characteristic features of the mass distributions of fission fragments in binary fission.

ASYMMETRICAL FISSION

The most characteristic feature of the fission of heavy nuclei at low excitation energies is the clearly pronounced asymmetry, i.e., the preferential fission into two fragments of substantial unequal mass.

Figures 2-5 show curves of the mass distributions of fragments for some typical instances of binary fission. Plotted on a log scale as ordinate is the total yield of a fragment of given mass, Y_i, expressed in percentages (while, in Fig. 5, the variables are proportional to the yields N_f), while the mass numbers A are plotted as abscissa.

As we see clearly in these diagrams, the asymmetry of the fission is manifested in the presence of two sharply defined maxima corresponding to mass numbers differing by approximately 40 to 50 mass units. Outside the region of the peaks, the yields decline abruptly. For example, the yield corresponding to symmetrical fission (i.e., fission into two equal fragments) in thermal fission of U^{235} is less than the yields in asymmetrical fission (i.e., into fragments corresponding to the peaks on the curves) by a factor of almost 600.

The position of the maxima depends on the mass number of the fissioning intermediate nucleus A_0 and on the excitation energy. In our earlier paper [30], formulas

$$\left. \begin{aligned} \overline{A}_H &= \frac{\sum\limits_{i} A_i Y_i}{\sum\limits_{i} Y_i} \ , \\[2em] \overline{A}_T &= \frac{\sum\limits_{k} A_k Y_k}{\sum\limits_{k} Y_k} \end{aligned} \right\} \tag{3}$$

were used for computing the average masses of the heavy $(\overline{A}_H)$ and light $(\overline{A}_L)$ fragments, respectively, characterizing the position of the maxima alluded to above. Y_k and Y_i are symbols for the respective yields.

From the yield determination we then find:

$$\sum\limits_{i} Y_i = \sum\limits_{k} Y_k = 100\%. \tag{4}$$

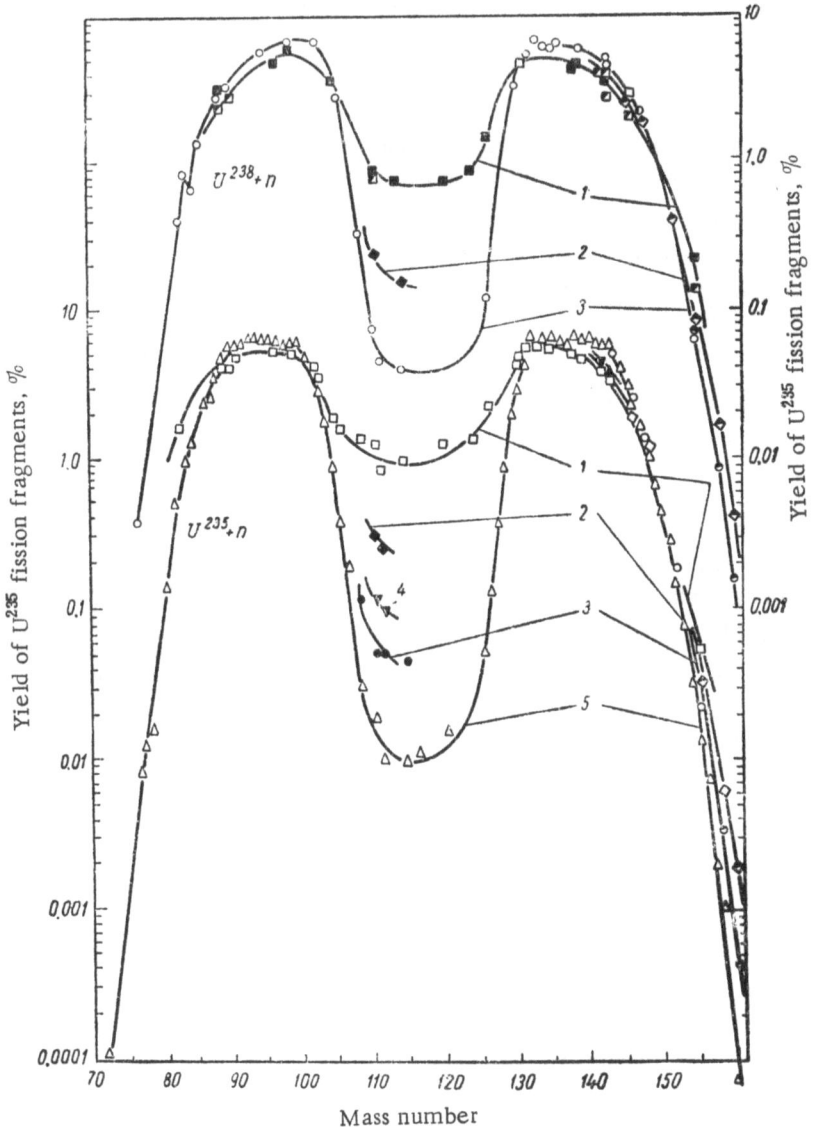

Fig. 2. Distribution curves for yields of fragments of fission of U²³⁵ and U²³⁸ by neutrons of various energies [27]: 1) 14 MeV neutrons; 2) 8 MeV neutrons; 3) fission spectrum neutrons; 4) 5 MeV neutrons; 5) thermal neutrons.

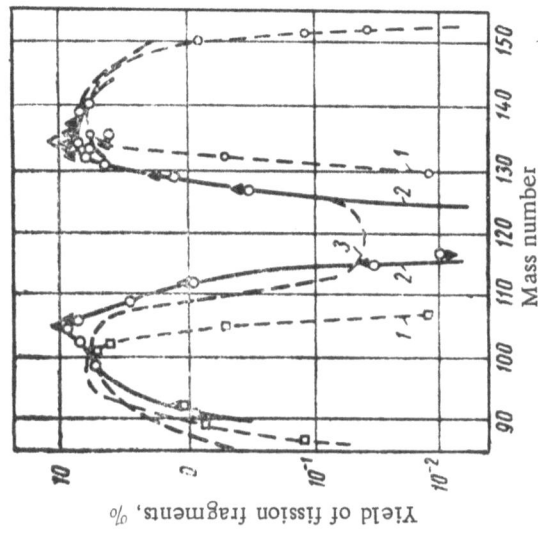

Fig. 4. Curves of distribution of fragment yields for spontaneous fission of U^{238} (1), Cm^{242} (2), and for fission of Pu^{239} (3) by slow neutrons [29].

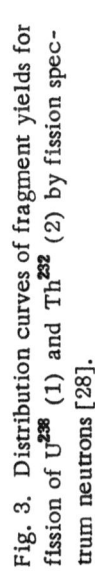

Fig. 3. Distribution curves of fragment yields for fission of U^{238} (1) and Th^{232} (2) by fission spectrum neutrons [28].

TABLE 1. Average Masses of Light and Heavy Fragments for Various Cases of Fission and for Computing $\bar{\nu}$

Fissionable isotope	Type of radiation inducing fission	Energy	A_0	$\bar{A}_L$	$\bar{A}_H$	$\bar{\nu} = A_0 - (\bar{A}_H + \bar{A}_L)$
U[233]	Neutrons	Thermal	234	93.3±0.1	138.2±0.1	2.5±0.2
U[235]	"	"	236	94.8±0.1	138.8±0.1	2.4±0.2
Pu[239]	"	"	240	98.5±0.2	139.0±0.1	2.5±0.3
Am[241]	"	Pile spectrum	242	101±0.3	138.3±0.3	·2.7±0.6
Th[232]	"	Fission spectrum	233	91.4±0.1	139.8±0.3	1.8±0.4
U[235]	"	Same	236	95.6±0.2	138.4±0.3	2.0±0.5
U[238]	"	"	239	97.3±0.1	138.9±0.1	2.8±0.2
U[235]	"	14 MeV	236	96.3±0.2	135.5±0.2	4.2±0.4
U[238]	"	14 MeV	239	97.0±0.3	137.0±0.4	5.0±0.4
U[238]	γ-Rays	8--10 MeV	238	96.6±0.1	138.7±0.3	2.7±0.4
U[238]	"	16 MeV	238	97.4±0.1	137.5±0.3	3.1±0.4
U[238]	"	48 MeV	238	96.5±0.2	137.2±0.3	4.3±0.5
Cf[252]	Spontaneous fission	—	252	106.3±0.3	140.9±0.3	4.8±0.6

Note, by the way, that the average number ν of secondary neutrons may be determined as the difference of the mass numbers [30]:

$$\bar{\nu} = A_0 - (\bar{A}_H + \bar{A}_L). \qquad (5)$$

All of these data are listed above in Table 1 for various cases.

According to these data, the maxima draw closer as A_0 increases, i.e., the degree of asymmetry tapers off.

Hence [31],

$$\bar{A}_H - \bar{A}_L = 288 - 1.04A_0 + \delta, \qquad (6)$$

where

$$\delta = \begin{cases} 0 & \text{for even } A_0 \\ 2 & \text{for odd } A_0. \end{cases}$$

$\bar{A}_L/\bar{A}_H$ and A_0 may be related by the empirical formula (Fig. 6):

$$\overline{A}_L/\overline{A}_H \approx 0.00556 A_0 - 0.625. \qquad (7)$$

Of course, only those fission events should be considered which show approximately identical excitation energies (spontaneous fission, thermal fission). Fission induced by fission spectrum neutrons was used as an exception.

It is typical that the approach of the maxima is primarily due to the increase in $\overline{A}_L$, since $\overline{A}_H$ remains virtually unaffected. Figure 7 displays the corresponding plots [32], the data points being plotted for various cases of fission at low excitation energies. The displacement of the heavy-fragment peak in the direction of larger masses begins to stand out in the fission of Cf^{252}.

The degree of observable asymmetry may be characterized by the parameter $a = \left(\dfrac{\overline{A}_H - \overline{A}_L}{A_0} \right)^2$. An empirical relationship has been found linking this parameter a to the parameter Z^2/A_0, which characterizes the fissioning nucleus (Fig. 8) [33]. This relationship is expressed by the formula

$$a = 81 \cdot 10^{-4} \left[(40.2 \pm 0.7) - \frac{Z^2}{A_0} \right], \qquad (8)$$

from which we learn that, when $Z^2/A_0 \approx 40.2$, asymmetry must tend to zero in fission.

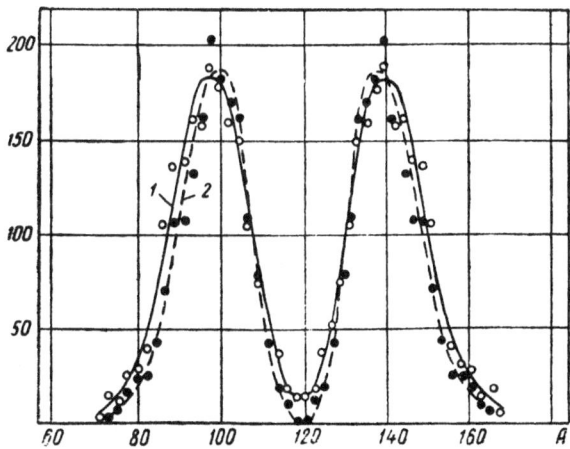

Fig. 5. Distribution curves of fission fragments yields [23]: 1) fission of Pu^{239} by thermal neutrons; 2) spontaneous fission of Pu^{240}.

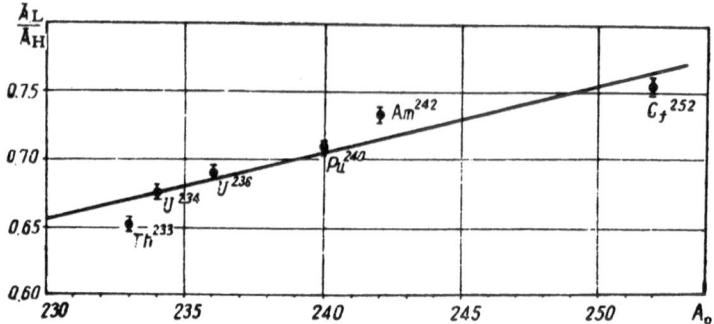

Fig. 6. Dependence of $\bar{A}_L/\bar{A}_H$ on A_0 (Cf^{252}—spontaneous fission; Th^{232}, U^{233}, U^{235}, Pu^{239}, Am^{241}—neutron-induced fission).

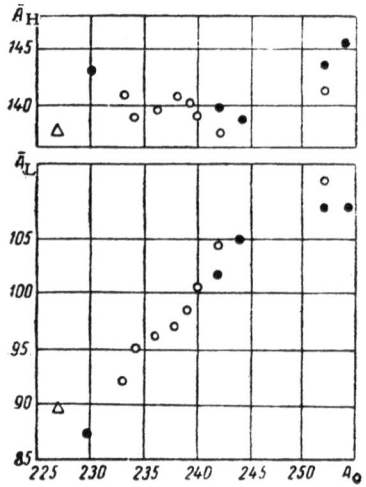

Fig. 7. Dependence of $\bar{A}_L$ and $\bar{A}_H$ on A_0 according to data of various authors [32].

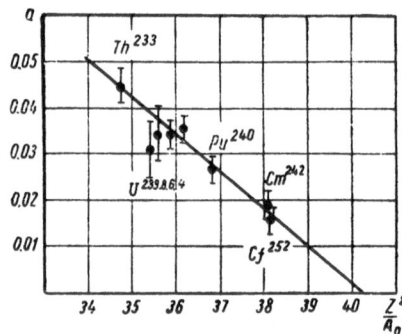

Fig. 8. Dependence of parameter $a = \left(\dfrac{\bar{A}_H - \bar{A}_L}{A_0} \right)^2$ which characterizes the degree of asymmetry, on Z^2/A_0 [33].

The relation (7) fails to satisfy all nuclides, however; a reciprocal trend is manifested in the case of different isotopes of the same element (e.g., U^{233}, U^{235}, and U^{238}), an outcome of the fact that $\bar{A}_H$ is approximately constant for those isotopes.

When we consider the mass distributions of fission fragments produced in the fission of U^{233}, U^{235}, Pu^{239} by thermal neutrons, Th^{232}, U^{238} by fission spectrum neutrons, and in the spontaneous fission of Cf^{252}, etc., our attention is drawn to the fact that the position of the left slope of the peak correspond-

TABLE 2. Fine Structure of Mass Distributions of Fission Fragments

Fissionable isotope	Average number of prompt neutrons	Position of peaks in fine structure (mass numbers)	
		Region of light fragments	Region of heavy fragments
U^{234}	2.5	99	133—134
U^{236}	2.5	100	134
U^{239}	2.7	103	134
Pu^{240}	3.0	100	134
Cm^{242}	3.0	105	134
Cf^{252}	3.8	113—114	134—135

ing to the heavy fragments remains unchanged on the curve, whereas the width of the distributions increases as the mass of the fissioning nucleus.

It is important to note that the distribution of the masses of heavy and light fragments becomes narrow in spontaneous fission of Th^{232}, U^{238}, Cm^{242}, and Cf^{252}, while the probability of symmetrical fission is lower than in fission of U^{233}, U^{235}, Pu^{239}, etc., by thermal neutrons.

For example, a direct comparison was made of the mass distribution in spontaneous and induced fission of the same nuclide, viz., Pu^{240}, which has a fairly short spontaneous fission period [23]; Pu^{240} is also an intermediate nucleus in the fission of Pu^{239} by thermal neutrons. As we realize from comparing the distribution curves (cf., Fig. 5), the peaks of the mass distribution curve in the case of spontaneous fission are narrower and the valley between them is deeper.

The asymmetry in the shape of the curve may be accounted for by shell effects, apparently, so that the preferential formation of fragments with closed shells of 50 protons and 82 neutrons in the fission process is an important factor. This is manifested in the region of mass numbers 127 to 139, which coincide with the left slope of the heavy-fragment peak on the curve. The shape of the curve is also related to some extent to the irregularity with which neutrons are emitted by fragments of different mass.

FINE STRUCTURE

One typical feature of the mass distribution of fragments in spontaneous fission (Cm^{242}, Cf^{252}) and in thermal fission (U^{233}, U^{235}, Pu^{239}) is the presence of a fine structure (Fig. 9) [1, 4, 16, 17, 28, 34-39, 142].

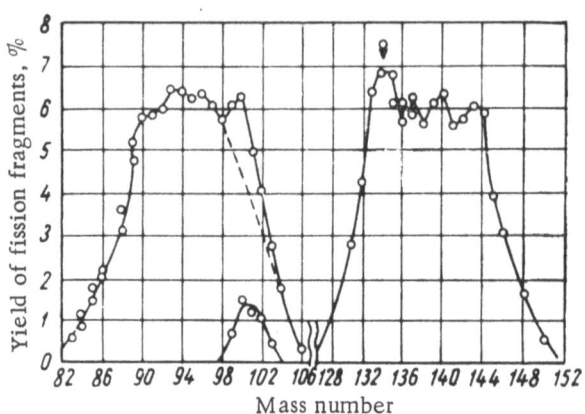

Fig. 9. Fine structure in the distribution curve of fragment yields in the thermal fission of U^{235} [4].

Even in earlier contributions to the literature [34-36], mass spectrometric methods, with their higher precision, successfully revealed deviations
from a smooth curve on the part of the relative yields of xenon and krypton
isotopes. The enhanced yield of fragments in the 133-135 mass region and
the yield of light fragments supplementary to it is observed in all the cases
enumerated above (see Table 2).

All of these data allow us to infer the preferential formation of fragments with a closed shell of 82 neutrons [4, 16, 17]. We might suppose, however, that the fine structure is due not only to the formation of nuclides with
closed shells and a "magic" number of neutrons (50, 82), but also to the corresponding enhanced probability of emission of prompt neutrons in the case
of fragments corresponding to certain mass numbers (this probability is also
due to the presence of closed shells in those nuclides).

As experiments have shown the fine structure is retained even when the
energy of excitation of the fissioning nuclide rises to 20 MeV. The height of
the fine-structure peaks then declines gradually [28].

EFFECT OF EXCITATION ENERGY ON FISSION OF VARIOUS NUCLEI.

SYMMETRICAL FISSION

It is clear from the totality of available experimental data that the
shape of the mass distribution curve of fission products depends on the energy
of excitation and on the atomic number Z of the fissioning nuclide.

As the energy of excitation increases, the shape of the distribution curve
varies substantially in the heavy-isotope region: the "valley" between the

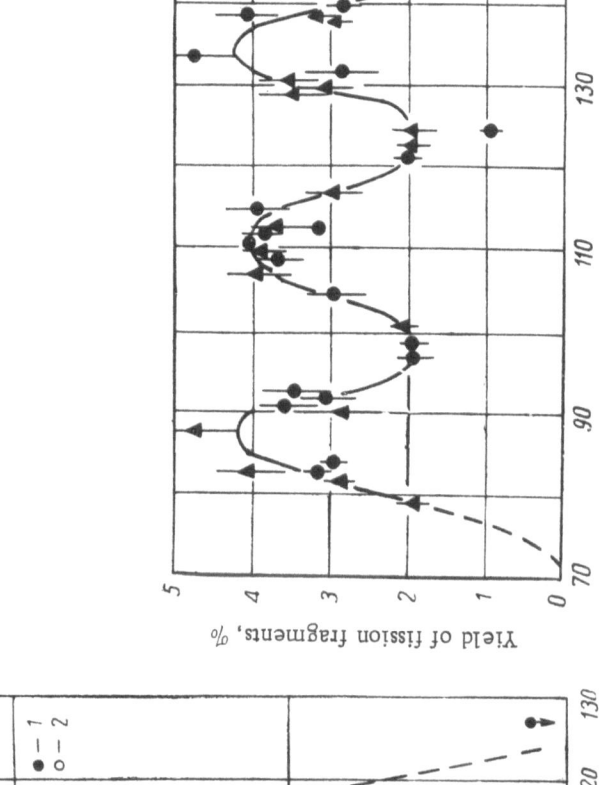

Fig. 11. Distribution curve of fission fragment yields in fission of Ra^{226} by 11 MeV protons. The compound nucleus Ac^{227} has an excitation of about 16 MeV [27, 41]: 1) measured values; 2) "reflected" values ($\nu = 5$).

Fig. 10. Distribution curve of yields of fission fragments of Bi^{209} fissioned by 22 MeV deutrons [40]: 1) reflected values; 2) "reflected" values ($\nu = 4$).

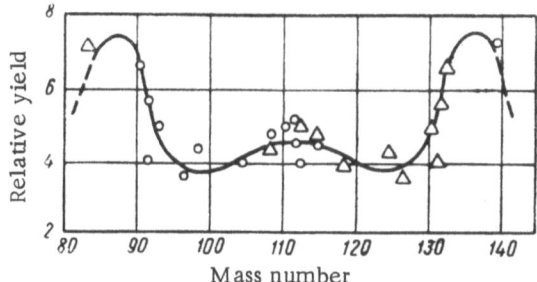

Fig. 12. Distribution curve of fragment yields in the fission of Ra^{226} by helium ions of 31 MeV energy. The excitation of the compound nucleus Th^{230} is roughly 26 MeV [28].

peaks characteristic of asymmetrical fission narrows. Simultaneously, the width of the mass distributions of the light and heavy isotopes increases (cf. Fig. 2).

At high excitation energies, there is in general a single maximum, and the fission exhibits an explicitly symmetrical character. This follows most obviously from inspection of the distribution curve of the fragments in the fission of Bi^{209} by 22 MeV deuterons (Fig. 10). We see from the curve that the probability of fission into fragments of closely equal masses has the highest value.

Figure 11 showed a most peculiar distribution obtained in the fission of Ra^{226} by 11 MeV protons. This distribution seems to be a superposition of two modes of fission—asymmetrical fission characterized by two extreme maxima

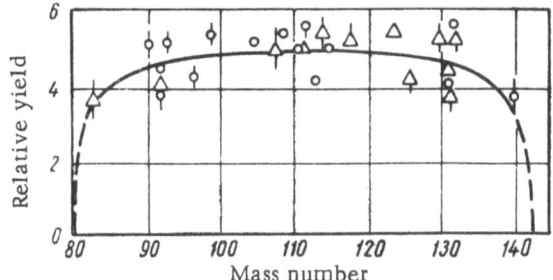

Fig. 13. Distribution curve of fragment yields in the fission of Ra^{226} by helium ions of 43 MeV energy. The excitation of the compound nucleus Th^{230} is about 38 MeV [28].

18

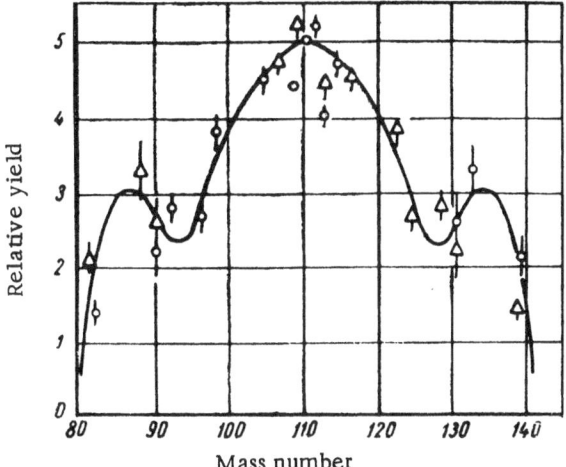

Fig. 14. Distribution curve of fragment yields in the fission of Ra^{226} by 21.5 MeV deuterons: the compound nucleus Ac^{228} has an excitation of 29.3 MeV [28].

and typical of the heavy nuclides (i.e., heavier than Ra^{226}), and symmetrical fission expressed by a peak at $\sim A_0/2$. In that base, both modes of fission have more or less the same probability of occurring. The two modes of fission are also observed in Th^{232} in photofission by gamma-photons such that $E_{max} = 70$ MeV [42].

At high excitation energies, the probability of symmetrical fission starts to predominate. However, in view of the fact that both of the peaks of asymmetrical fission and the peak of symmetrical fission expand still further as the energy is increased, there is a superposition of all the maxima, and the resulting distribution curve does not necessarily feature three maxima, as we see in the case of fission of Ra^{226} by 11 MeV protons (Figs. 12-15). We may apparently surmise [28] that two independent modes of fission, asymmetrical and symmetrical, are manifested in the fission of heavy elements.

An attempt has been made [43] to account for the abrupt change in the nature of fission symmetry in the region $82 \leq Z \leq 90$ (Pb—Th). Whereas lead and bismuth fission symmetrically at excitations of the order of 10 to 30 MeV, thorium and all the heavier elements fission asymmetrically. Radium ($Z = 88$) occupies an intermediate position with respect to the atomic number, and reveals an intermediate fission slope (cf. Fig. 11). At the same time, attention has been given [43] to the fact that an abrupt change in the extent to which the core of the nucleus is affected by nucleons outside the closed shells (with

19

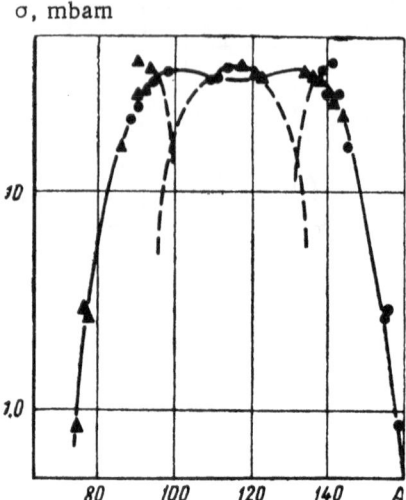

Fig. 15. Distribution curve of fragment yields in the fission of U^{235} by helium ions of 32.8 energy; the unbroken curve is plotted from the experimentally derived data. The broken-line curve shows the breakdown of the experimental curve into two components characterizing the symmetrical and asymmetrical modes of fission [28].

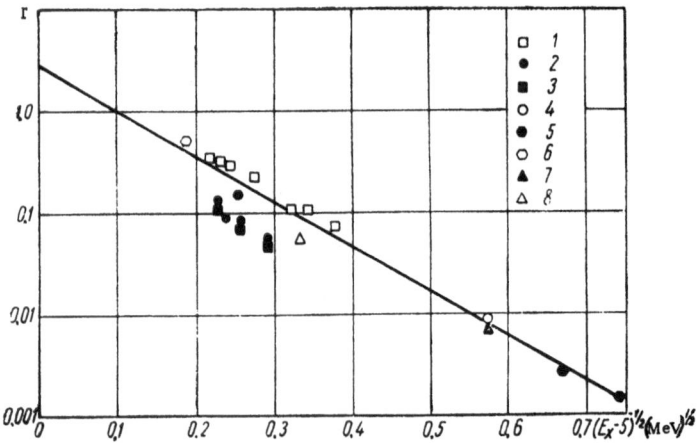

Fig. 16. Symmetry of fission as a function of the energy of excitation, according to data reported by various authors [45]: 1) $Th^{232} + p$; 2) $U^{238} + p$; 3) $U^{235} + p$; 4) $U^{238} + n$; 5) $U^{235} + n$; 6) $Th^{232} + \alpha$; 7) $Th^{232} + n$; 8) $U^{235} + \gamma$.

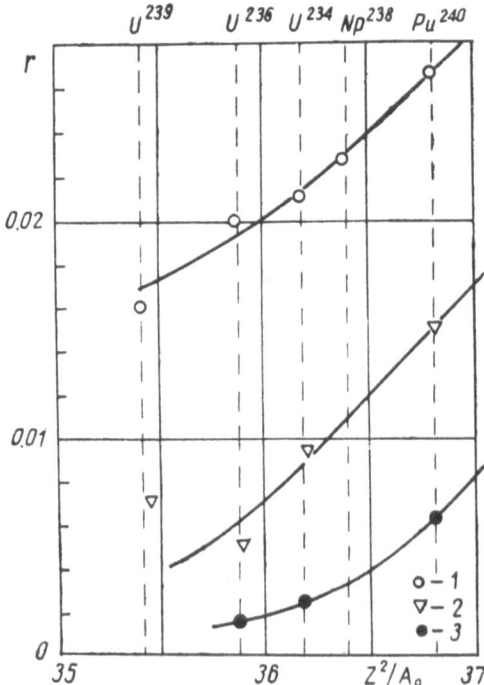

Fig. 17. Dependence of the relative probability of symmetrical fission r on the ratio Z^2/A_0. Neutron-induced fission [31]: 1) by 14.5 MeV neutrons (values plotted on ordinate scale have been reduced to one-tenth); 2) fission spectrum neutrons; 3) thermal neutrons.

proton number 82 and neutron number 126). It is a natural assumption that the outer nucleons and the number of such nucleons are capable of drastically affecting the equilibrium shape of the nucleus, may exert a vigorous effect on the process of deformation of the nucleus, and consequently on the nature of the fission symmetry. The fission will be either symmetrical or asymmetrical depending on how many nucleons are required to form the closed shells or on what sort of neutron excess exists over and above the closed shells. Further experiments will be needed in order to develop this point of view on the symmetry of fission.

The following must also be noted: as the excitation energy is increased, the number of secondary neutrons also increases, and consequently, according to formula (5), the two peaks of asymmetrical fission approach (cf. Table 1).

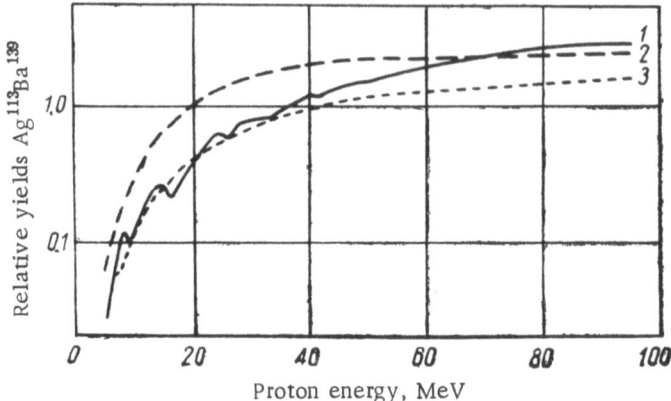

Fig. 18. Dependence of the ratio of yields of Ba^{139} and Ag^{113} on proton energy in the fission of Th^{232} (1), Pu^{239} (2), and U^{238} (3)[46].

In the fission of various nuclides by neutrons of the same energy, the position of the mass peak of heavy elements remains practically unchanged[44].

The approach of the two peaks, mentioned above, and the predominance of symmetrical fission at excitation energies above 40 MeV may be related to the diminishing influence of shell effects as the excitation energy increases.

The relationship between the two modes of fission may be characterized by the ratio $r = Y_{min}/Y_{max}$, which acts as a rather crude characteristic of this relationship, since the distribution observed experimentally is quite complex and the yields at the maxima Y_{max} and at the minimum Y_{min} are not determined solely by the probabilities of the asymmetrical and symmetrical modes, respectively, but are in each case the totality of the probabilities of fission in each mode.

Figure 16 shows the relationship between the ratio r for various cases of fission and the parameter $(E_x - 5)^{-1/2}$ (MeV)$^{-1/2}$, which typifies the energy of excitation ($\sim$5 MeV – the energy available for the critical deformation of the nucleus at the saddle point)[45]. Clearly, from the drawing, the increase in the energy of excitation of the compound nucleus is accompanied by an increase in the relative probability of symmetrical fission.

Aside from this quite familiar dependence of the ratio r on the energy of excitation of the compound nucleus, there is also observed a change in r related to the value of the ratio Z^2/A_0 of the fissioning nucleus[31, 44]. This relationship is shown in Fig. 17 for cases of neutron-induced fission of U^{238}, U^{235}, U^{233}, Np^{237}, Pu^{239}.

One of the reasons for the increase in r as Z^2/A_0 increases is apparently the increase in the excess energy of excitation of the fissioning nuclide over and above the fission barrier at the same energy of the neutrons inducing fission.

A detailed investigation has been made of the dependence of the ratios of Ag^{113} and Ba^{139} yields (characterizing the probability of symmetrical fission) on the energy of protons responsible for fission of Th^{232}, U^{238}, Pu^{239} [46]. Measurements of the yields were carried out over the range of energies extending to 100 MeV. The interval between experimental points on the first half of the curves amounts to approximately 1 MeV. The results of these researches appear in Fig. 18.

The presence of periodically repeating dips in the curve is rather conspicuous. The position of the dips corresponds to evaporation, prior to the onset of fission, of one, two, or more neutrons, and accordingly to a certain cooling of the fissioning nucleus and a concomitant increased probability of asymmetrical fission.

QUANTUM CHARACTERISTICS OF THE FISSIONING NUCLEUS
AND FISSION ASYMMETRY

Some theoretical arguments have been expressed to the effect that the probability of symmetrical fission is dependent on the spin state of those fission channels which make the principal contribution to fission [47, 48].

Consider an example. The spin and parity of the nuclide U^{235} in the ground state is $7^-/2$. The intermediate nucleus will be in the 3^- state or the 4^- state after capturing a slow neutron. The assumption is that the mass distribution of the fragments in fission of the intermediate nucleus U^{236} in the 3^- state by neutrons of close to thermal energies will differ from the distribution obtained for the U^{236} nucleus in the 4^- state.

In order to determine the effect of quantum characteristics on the asymmetry of fission, several investigations were made of U^{233} [22, 49, 50], U^{235} [22, 51, 52], Pu^{239} [53], Pu^{241} [53], and Am^{241} [54]. Different measurements failed to produce harmonious results on the dependence of the relationship linking the symmetrical and asymmetrical modes of fission in the case of fission by thermal neutrons and by resonance neutrons, and this question cannot be viewed as clarified satisfactorily at this time.

THEORETICAL ELUCIDATION OF ASYMMETRY IN FISSION

A large number of theoretical papers has appeared in the literature in the recent period in which the authors have attempted to account for the fundamental feature of the mass distribution of fission fragments— asymmetry of fission—in some satisfactory manner [55]. However, as yet there is still no complete theoretical explanation of this asymmetry. The fundamental aspects of various theoretical investigations in this direction are enumerated below.

The collective motions of nucleons underlying the fission process may be described, in a first approximation, by the simple model of a drop of a homogeneous almost incompressible fluid. Because of the low compressibility of nuclear matter, the internal collective motions of the nuclei in this model lead to an alteration of the form of the nuclear surface.

The asymmetry in fission has not met with any explanation drawing upon the original variant of the theory of fission provided by Bohr, Wheeler, and Frenkel, based on the liquid drop model [56-60], since the minimum value of the fission barrier corresponds to symmetrical fission. It is true that an attempt was later made, within the framework of Frenkel's theory [61], to account for the asymmetry in terms of the decisive influence of subbarrier fission. According to these concepts, the energy of excitation is initially distributed, over a protracted span of time, between many possible competing modes of oscillation of the liquid drop nucleus, and fission cannot take place until the oscillations acquire sufficiently high amplitudes. Fission by tunneling underneath the potential barrier then has its highest probability of occurrence at certain mass values of fragments not equal to each other, since the reduced mass of the fragments is included in the expression for the probability of fission underneath the barrier. However, quantitative estimates have failed to confirm this theory. No success has been had either in providing a satisfactory explanation for the asymmetry by invoking variants of the liquid drop model in which compressibility, inhomogeneity, and polarizability of the nuclear matter, as well as various dynamical effects, are taken into account [62-68].

The extraordinary complexity of the fission process has so far defied attempts to combine the complete description of the entire array of experimental facts describing asymmetry into a single unified theory. In particular, Fong's theory [69, 70] assumes that the fission probability is determined solely by the statistical weight of the final state of the fragments, which increases drastically as the energy of excitation of the fragments. (The energy of excitation is found from the semi-empirical formula for the binding energy of nuclides, and corrections taking shell effects into account have been introduced into the formula.) Fong derived, theoretically, a curve for the mass distribution of U^{235} fission products in thermal fission, satisfactorily in accord with the experimentally derived curve. However, no satisfactory fit with the experimental data was achieved in the case of thermal fission of Pu^{239} [71] nor in several variants of the statistical theory.

The imperfections of this theory are due to a number of simplifying assumptions, particularly the neglect of the fact that the density of the states at equal energies of excitation will be less in the case of magic-number fragments than in the case of fragments not associated with magic numbers.

B. T. Geilikman [72-74] computed the energy of the fissioning nucleus before the rupture of the "neck," taking due account of shell effects, in his studies of the mass distribution of fragments. The approximate nature of the calculations (quadratic approximation in computing the energy of deformation and the energy of the Coulomb deformation), the inaccuracy of the Weizsäcker formula with the shell effects taken into account, combined in this case, however, to thwart satisfactory agreement with experimental data.

Note, finally, that the weak dependence of $\bar{A}_H$ on $\bar{A}_L$ is indicative of the substantial influence exerted by the shell structure of a heavy fragment on fission asymmetry. In the papers where due account was taken of this point, however, the asphericity of the shape of the fissioning nuclei was ignored, and this factor is apparently also crucial in the asymmetry of fission [73].

TABLES OF FISSION FRAGMENT YIELDS

The tables below include data on the yields of fission fragments referable to various cases of fission at excitation energies extending to 100 MeV,
compiled over the period from 1939 to 1962.

Tables 3 to 5 include yields of fragments in the spontaneous fission of
Th^{232}, U^{238}, Cm^{242}, and Cf^{252}. The dependence of the yields of krypton and
xenon isotopes in the spontaneous fission of uranium in ores on the concentration of those isotopes in uranium is shown (Table 4) to be a consequence
of the fission of U^{235} by slow neutrons.

Tables 6 and 7 present fragment yields in the fission of U^{233}, U^{235}, Pu^{239},
Am^{241}, and Cm^{242} by thermal neutrons.

Tables 8 to 19 contain fragment yields in the fission of heavy nuclides
by neutrons as a function of the increase in the energy of excitation (resonance neutrons in Tables 8 to 11, fission spectrum neutrons in Table 13, neutrons of ~8 MeV energy in Table 12, neutrons of 14.5 MeV energy in Table
14). Tables 18 and 20 present the independent and the cumulative yields for
the cases of fission by thermal neutrons and by 14 MeV neutrons.

Tables 21 and 22 list yields of fission products of heavy nuclei for fission by gamma rays of various energies, while Tables 23 to 39 do the same
for fission by charged particles (Tables 23 to 25 for fission by protons, Tables
19, 26 to 28 for deuterons, Tables 29 to 38 for helium nuclei, Table 39 for
C^{12} ions).

The number of isotopes for which yields have been determined in various cases of fission by neutrons extends over quite a wide variety. The larg-
est number of isotopes for which yields have been determined is associated
with the case of fission of U^{235} by thermal neutrons (over 275 isotopes). All
of the isotopes of a single element are grouped together.

In Tables 3 through 39, some of the figures are tabulated in brackets,
as they appeared in the original versions. These figures are either figures obtained by indirect means, or less reliable figures, or else results of estimates.
Some of the papers [75-79, 140, 150] presented the yields in the form of graphs.
The data from these papers are not reproduced in this book.

TABLE 3. Yields of Fragments in Spontaneous Fission of Th^{232}, U^{238}, Cm^{242}, and Cf^{252} (%)

Nu-clide	Half-life	Th^{232} [81]	U^{238}* [81—84, 143]	Cm^{242}† [39]	Cf^{252}† [85—86]
Kr^{83}	Stable	0.036±0.025	0.036±0.015 [81] / 0.12±0.01 [82]	—	—
Kr^{84}	"	0.180±0.040	0.119±0.040 [81] / 0.45±0.05 [82]	—	—
Kr^{86}	"	0.87±0.12	0.75±0.11 [81] / 1.64±0.15 [82]	—	—
Sr^{89}	50.5 days	—	5.9±1.4 / 2.9±0.3 } [143]	—	—
Sr^{90}	27.7 yr	—	6.8±0.6 [143]	—	—
Sr^{91}	9.67 h	—	5±4 / 6.9±0.5 } [143]	0.95±0.3	—
Sr^{92}	2.60 h	—	11±4 [143]	1.2±0.3	—
Zr^{97}	17.0 h	—	—	—	2.1±0.3 [86]
Mo^{99}	66.0 h	—	6.4 [83] / 6.3±0.6 / 6.0±0.5 } [143]	5.7±0.7	2.2±0.5 [85] / 3.0±0.45 [86]
Mo^{101}	14.61 min	—	—	—	4.1±0.8 [86]
Ru^{103}	39.8 min	—	—	7.2±1.5	—

• Yields of isotopes in spontaneous fission of uranium in ores depend on the uranium concentration in those ores (because of the additional contribution due to the effect of fission of U^{235} by slow neutrons) [80].

† These figures indicate the total yield of the chain with a given mass number A.

TABLE 3 (Continued)

Nuclide	Half-life	Th232 [81]	U238* [81—84, 143]	Cm242† [39]	Cf252† [85, 86]
Ru105	4.5 h	—	—	9.9±1.0	9.2±1.4 [85]
Ru106	1.00 yr	—	—	8.4±1.0	—
Pd109	13.5 h	—	< 0.02 [143]	2.9±0.4	6.8±1.3 [85]
Pd112	21 h	—	—	1.1±0.2	4.5±0.9 ⎫ 4.8±1.0 ⎬ [85]
Ag111	7.6 days	—	< 0.05 [143]	—	4.5±0.9 [85]
Ag113	5.3 h	—	—	—	4.2±0.8 [85]
Cd115m	43 days	—	—	(0.003)	—
Cd115	53 h	—	< 0.05 [143]	0.033±0.01	2.8±0.5 [85]
Cd115 (Total)	—	—	—	0.036±0.01	—
Cd117m	3.0 h	—	—	< 0.01	—
In117	~50 min	—	—	—	≤ 1.0 [85]
Sb127	88 h	—	—	0.37±0.1	—
Sb129	4.2 h	—	—	1.7±0.4	—
Te131m	30 h	—	—	2.3±0.5	—
Te132	77.7 h	—	4.5±0.5 [143]	7.4±1.3	2.8±0.4 ⎫ 3.5±0.5 ⎬ [85]

	Half-life				
I131	8.08 days	—	0.42 ± 0.14 [143]	4.3 ± 0.7	—
I132	2.26 h	—	3.47 ± 0.42 [143]	—	—
I133	20.8 h	—	1.4 ± 0.3 [143]	6.0 ± 0.9	4.8 ± 0.7 / 5.1 ± 0.8 } [85]
I134	52.5 min	—	5.0 ± 0.6 [143]	8.0 ± 1.3	4.2 ± 0.6 / 4.8 ± 0.7 } [85]
I135	6.68 h	—	4.9 ± 0.6 [143]	7.3 ± 1.4	4.0 ± 0.6 / 5.1 ± 0.8 } [85]
Xe129	Stable	0	< 0.012 [81] / 0.088 ± 0.013 [82]	—	—
Xe131	"	0.509 ± 0.02	0.455 ± 0.02 [81] / 0.74 ± 0.02 [82]	—	—
Xe132	"	3.63 ± 0.08	3.46 ± 0.025 [82] / 3.57 ± 0.06 [81]	—	—
Xe134	"	5.12 ± 0.10	4.99 ± 0.07 [81] / 5.10 ± 0.014 [82]	—	—
Xe136	"	6.0	6.00 [82] / 6.00 [81]	—	—
Cs136	12.9 days	—	—	0.80 ± 0.12	—
Cs138	32.2 min	—	—	—	6.3 ± 0.9 [85]
Ba139	84.0 min	—	—	6.6 ± 0.7	6.2 ± 0.9 [85]
Ba140	12.80 days	—	6.1 [84] / 9.6 ± 1.2 [143]	5.9 ± 0.8	—
Ce143	33 h	—	7.9 ± 1.4 [143]	—	7.8 ± 1.5 [87]
Ce144	285 days	—	6.5 ± 0.5 [143]	—	—
Pr143	13.76 days	—	7.5 ± 0.5 [143]	—	7.4 ± 1.5 [86]
Nd147	11 06 days	—	4.2 ± 0.4 [143]	—	4.0 ± 0.8 [86]
Sm153	47.1 h	—	—	—	1.3 ± 0.3 [86]

TABLE 4. Dependence of Yields of Krypton and Xenon Isotopes in Spotaneous Fission and in Neutron-Induced Fission of Uranium in Ores on the Concentration of Uranium in Those Ores [80]

| Nuclide | Half-life | Yield of isotopes, % | | | | | | From uraninite containing 60% U (Cardiff township, Ontario, Canada) |
| | | from pitchblende containing U_3O_8 | | | | | | |
		65.21% Katanga, Congo	45.5% Eagle Mine Beaverlodge, Canada	16.95% Ice Mine Beaverlodge, Canada	13.99% Lake Athabaska, Canada	27.55% Nesbitt-Labine Beaverlodge, Canada	36.46% Great Bear Lake, Canada	
Kr^{83}	Stable	0.30±0.06	—	—	—	—	0.073±0.022	—
Kr^{84}	"	0.50±0.13	—	—	—	—	0.34±0.06	—
Kr^{86}	"	1.62±0.12	—	—	—	—	1.19±0.05	—
Xe^{129}	"	0.217±0.036	0.223±0.037	0.226±0.035	0.093±0.042	0.107±0.060	0.103±0.062	0.029±0.002
Xe^{131}	"	1.27±0.04	1.16±0.04	1.07±0.04	0.799±0.040	0.830±0.059	0.798±0.054	0.613±0.010
Xe^{132}	"	3.91±0.06	3.91±0.06	3.89±0.07	3.79±0.07	3.80±0.06	3.73±0.08	3.74±0.03
Xe^{134}	"	6.03±0.03	5.93±0.03	5.79±0.06	5.58±0.04	5.56±0.05	5.54±0.6	5.41±0.03
Xe^{136}	"	6.50	6.50	6.50	6.50	6.50	6.50	6.50

TABLE 5. Independent Yields of Fragments in Spontaneous Fission of Cm^{242}, Independent and Cumulative Yields of Fission Fragments of Cf^{252} (ratios to total yields of chains) [147]

Nuclide	Half-life	Cm^{242}	Cf^{252} Independent	Cf^{252} Cumulative
Cs^{136}	12.9 days	0.12±0.02	0.008	—
Xe^{139}	41 sec	—	—	0.67±0.01
Xe^{140}	16 sec	—	—	0.45±0.01
Xe^{141}	1.7 sec	—	—	0.172±0.005
Xe^{144}	1 sec	—	—	< 0.007

31

2. NEUTRON-INDUCED FISSION

TABLE 6. Fragment Yields in Thermal Fission of U^{233}, U^{235}, Pu^{239}, Am^{241}, Cm^{242} (%)

Nuclide	Half-life	U^{233} [1, 3, 4, 6, 9, 13—15, 20, 26, 88*—90, 97, 104]	U^{235} [1—3, 4, 6, 14, 20, 38, 51, 80, 88*, 91*, 92*—97*, 104]	Pu^{239} [1, 3, 4, 6, 14, 20, 87†, 88*, 97*, 98, 99, 100—102, 104]	Am^{241}‡ [103]	Cm^{242} [88*]
Be^7	54 days	$< 3 \cdot 10^{-7}$ [26]	—		—	—
Mg^{28}	21 h	$< 4.2 \cdot 10^{-9}$ [26]	—		—	—
Ni^{66}	56 h	$(2.0 \pm 1.0) \cdot 10^{-8}$ [26]	—		—	—
Zn^{72}	49 h	—	$1.6 \cdot 10^{-5}$ [1,6] $1.5 \cdot 10^{-5}$ [2]	0.00012 [1, 6]	—	—
Zn^{73}	<2 min	—	$(9.8 \cdot 10^{-5})$ [2]	—	—	—
Ga^{72}	14.3 h	—	$1.5 \cdot 10^{-5}$ [2]	—	—	—
Ga^{73}	5.0 h	—	$1.1 \cdot 10^{-4}$ [1, 6] $1.0 \cdot 10^{-4}$ [2]	—	—	—
Ga^{74}	7.8 min	—	$3.5 \cdot 10^{-4}$ [6]	—	—	—
Ge^{75}	82 min	—	$8 \cdot 10^{-4}$ [2]	—	—	—
Ge^{77m}	54 sec	—	0.0054 [2]	—	—	—
Ge^{77}	11.3 h	0.010 [1] 0.011 [6]	0.0031 [1, 6] 0.0037 [2]	—	—	—
Ge^{78}	86 min	—	0.019 [1] 0.018 [2] 0.020 [6]	—	—	—

Isotope	Half-life					Yield
As77	38.7 h	—	—	0.019 [1] / 0.021 [6]	—	0.0083 [1, 6] / 0.0091 [2]
As78	91 min	—	—	—	—	0.020 [6] / 0.021 [1] / 0.02 [2] / 1.8·10⁻³ [88]
As79	9.0 min	—	—	—	—	0.056 [1, 6] / 0.04 [2]
As81	<10 min	—	—	—	—	(0.125) [2]
Se77m	17.5 sec	—	—	—	—	< 2·10⁻⁴ [2]
Se79m	3.91 min	—	—	—	—	0.04 [2]
Se79	≤6.5·10⁴ yr	—	—	—	—	0.04 [2]
Se81m	56.8 min	—	—	—	—	0.0084 [1, 6] / 0.008 [2]
Se81	18.2 min	—	—	—	—	0.14 [1, 6] / 0.133 [2]
Se83m	70 sec	—	—	—	—	0.30 [2]
Se83	25 min	—	—	—	—	0.22 [1, 6] / 0.21 [2]
Se84	3.3 min	—	—	—	—	1.1 [2]
Se85	33 sec	—	—	—	—	~1.1 [6]

*Independent yields are given in [88, 91, 92, 97, 104].

†Only relative yields appear in [87] (data referable to highly irradiated plutonium, viz., to 2.7·10²² neutrons/cm irradiation); burnup of the fragments occurred.

‡Fission by pile neutrons [103].

TABLE 6 (Continued)

Nuclide	Half-life	U²³³ [1, 3, 4, 6, 9, 13—15, 20, 26, 88*—90, 97, 104]	U²³⁵ [1—3, 4, 6, 14, 20, 38, 51, 80, 88*, 91*, 92*—97*, 104]	Pu²³⁹ [1, 3, 4, 6, 14, 20, 87 †, 88*, 97*, 98, 99, 100—102, 104]	Am²⁴¹ ‡ [103]	Cm²⁴² [88*]
Br⁸²	35.87 h	(7.46±0.17)·10⁻⁴ [26] 1.1·10⁻³ [6]	3.8·10⁻⁵ [2] 3.5·10⁻⁵ [88] 4·10⁻⁵ [6]	—	—	—
Br⁸³	2.30 h	0.79 [1] 0.87 [6]	0.51 [1, 6] 0.48 [2] 0.40 [3]	0.085 [1] 0.084 [6]	—	—
Br⁸⁴	6.0 min 31.8 min	— —	0.019 [1, 6] 0.90 [1] 0.92 [6] 1.1 [2]	—	—	—
Br⁸⁵	3.0 min	—	1.5 [2]	—	—	—
Br⁸⁷	55.6 sec	—	2.7 [2]	—	—	—
Br⁸⁸	15.5 sec	—	(2.9) [2]	—	—	—
Kr⁸³ᵐ	114 min	—	0.48 [2]	—	—	—
Kr⁸³	Stable	1.14 [1] 1.17 [6]	0.544 [1, 6] 0.67 [80]	0.29±0.01 [99] 0.29 [6]	—	—
Kr⁸⁴	"	1.90 [1] 1.95 [6]	1.00 [1, 6] 1.27 [80]	0.47±0.02 [99] 0.47 [6]	—	—
Kr⁸⁵ᵐ	4.36 h		1.5 [2]	—		—
Kr⁸⁵	10.3 yr	0.56 [1] 0.58 [6]	0.293 [1, 6] 0.3 [2]	0.55±0.03 [99] 0.127 [6]	—	—

Isotope	Half-life						
Kr⁸⁶	Stable	3.18 [1] 3.27 [6]	—	2.45 [80] (2.02) [1, 6]	0.75 ± 0.02 [99] 0.76 [6]	—	—
Kr⁸⁷	78 min		—	2.7 [2]		—	—
Kr⁸⁸	2.77 h		—	3.7 [2]		—	—
Kr⁸⁹	3.18 min		—	4.59 [1] (4.6) [2]		—	—
Kr⁹⁰	33 sec		—	5.0 [1] (5.2) [2]	—	—	—
Kr⁹¹	9.8 sec		—	3.45 [1] (3.7) [2]	—	—	—
Kr⁹²	3.0 sec		—	1.87 [1] (2.7) [2]	—	—	—
Kr⁹³	2.0 sec		—	(1.3) [2]	—	—	—
Kr⁹⁴	1.4 sec		—	(0.6) [2]	—	—	—
Kr⁹⁵	Short-lived		—	(0.2) [2]	—	—	—
Kr⁹⁷	~1 sec		—	(~0) [2]	—	—	—
Rb⁸⁵	Stable	2.51 [6]	—	1.30 [1, 6] 1.3 [38]	0.539 [6]	—	—
Rb⁸⁶	18.66 days	2.3·10⁻⁴ [6]	—	3.1·10⁻⁵ [88] 2.8·10⁻⁵ [2] 2.9·10⁻⁵ [6]	1.1·10⁻⁴ [88] 2.3·10⁻⁵ [6]	—	—
Rb⁸⁷	5.0·10¹⁰ yr	4.56 [6]	—	2.49 [1, 6] 2.75 [38] 2.7 [2]	0.92 [6]	—	—
Rb⁸⁸	17.8 min		—	3.7 [2]	—	—	—
Rb⁸⁹	15.4 min		—	4.8 [2]	—	—	—
Rb⁹⁰	2.74 min		—	5.9 [2]	—	—	—

35

TABLE 6 (Continued)

Nuclide	Half-life	U²³³ [1, 3, 4, 6, 9, 13—15, 20, 26, 88*—90, 97, 104]	U²³⁵ [1—3, 4, 6, 14, 20, 38, 51, 80, 88*, 91*, 92*—97*, 104]	Pu²³⁹ [1, 3, 4, 6, 14, 20, 87†, 88*, 97*, 98, 99, 100—102, 104]	Am²⁴¹† [103]	Cm²⁴² [88*]
Rb⁹¹	14 min	—	(5.7) [2]	—	—	—
Rb⁹²	80 sec	—	(5.5) [2]	—	—	—
Rb⁹³	Short-lived	—	(4.4) [2]	—	—	—
Rb⁹⁴	"	—	(2.9) [2]	—	—	—
Rb⁹⁵	"	—	(1.6) [2]	—	—	—
Rb⁹⁷	"	—	(0.1) [2]	—	—	—
Sr⁸⁸	>3·10¹⁶ yr	5.30±0.30 [14] 5.37 [6]	3.57 [1, 6]	1.39±0.04 [14] 1.42 [6]	—	—
Sr⁸⁹	50.5 days	6.5 [1] 5.86 [6] 6.3±0.3 [15] 5.56±0.15 [90]	4.79 [1, 6] 4.8 [2, 51]	1.9 [1] 1.71 [6]	0.81± ±0.05	—
Sr⁹⁰	27.7 yr	5.80±0.40 [14] 6.43 [6] 4.56±0.08 [89] 6.19±0.03 [90]	5.77 [1, 6] 5.9 [2]	2.25 [6] 2.31±0.05 [14]	—	—
Sr⁹¹	9.67 h	5.57 [6] 4.82±0.25 [90]	5.81 [1, 6] 5.9 [2]	2.43 [6] 2.4 [1]	—	—
Sr⁹²	2.60 h	—	5.3 [1,6] 6.1 [2]	—	—	—
Sr⁹³	8.2 min	—	(6.4) [2]	—	—	—

Isotope	Half-life					
Sr⁹⁴ → Sr^{94}	1.3 min	—	(5.8) [2]	—	—	—
Sr^{95}	~0.7 min	—	(4.7) [2]	—	—	—
Sr^{97}	Short-lived	—	(1.7) [2]	—	—	—
Y^{90}	64.2 h	< 4.0 · 10⁻⁴ [88]	5.77 [1] / <0.0013 [91] / 5.9 [2] / <4.0·10⁻⁴ [88]	—	—	—
Y^{91m}	50.3 min	—	2.4 [2]	—	—	—
Y^{91}	57.5 days	5.1 [6] / 4.1 [3] / 3.55±0.06 [90]	~5.4 [1,6] / 5.9 [2] / <0.05 [91]	3.0 [1] / 2.9 [6]	1.16± ±0.08	—
Y^{92}	3.60 h	—	6.1 [2]	—	—	—
Y^{93}	10.4 h	—	6.5 [2] / 6.1 [6]	—	—	—
Y^{94}	16.5 min	—	5.4 [1] / 6.5 [2]	—	—	—
Y^{95}	10.5 min	—	6.4 [2]	—	—	—
Y^{97}	Short-lived	—	(4.8) [2]	—	—	—
Zr^{91}	Stable	6.53 [1] / 6.43 [6]	5.84 [1,6]	2.61 [6]	—	—
Zr^{92}	"	6.70 [1] / 6.64 [6]	6.03 [1,6]	3.14 [6]	—	—
Zr^{93}	1.1 · 10⁶ yr	7.10 [1] / 6.98 [6]	6.45 [1,6] / 6.5 [2]	3.97 [6]	—	—
Zr^{94}	Stable	6.82 [1] / 6.68 [6]	6.40 [1,6]	4.48 [6]	—	—

TABLE 6 (Continued)

Nuclide	Half-life	U233 [1, 3, 4, 6, 9, 13—15, 20, 26, 88*—90, 97, 104]	U235 [1—3, 4, 6, 14, 20, 38, 51, 80, 88*, 91*, 92*—97*, 104]	Pu239 [1, 3, 4, 6, 14, 20, 87†, 88*, 97*, 98, 99, 100—102, 104]	Am241 ‡ [103]	Cm242 [88*]
Zr95	65 days	5.9 [1] 5.01±0,56 [90] 6.1 [6]	6.2 [1,6] 6.4 [2]	5.8 [6] 5.9 [1]	3.90± ±0.51	—
Zr96	>3.6·10^17 yr	5.60 [1] 5.58 [6]	6.33 [1,6]	5.17 [6]	—	—
Zr97	17.0 h	—	5.9 [1,6] 6.2 [2]	5.5 [6] 5.6 [1]	3.55± ±0.46	—
Nb93m	12 yr	—	2.1 [2]	—	—	—
Nb95m	90 h	—	0.06 [2]	—	—	—
Nb95	35 days	5.16±0.64 [90]	6.4 [2]	—	—	—
Nb96	23.35 h	(5.7±1.0)·10^-3 [97] 6.5·10^-3 [6]	5.7·10^-4 [2,88] (5.85±1.0)·10^-4 [97] 6.1·10^-4 [6]	(3.6±0.6)·10^-3 [97] 3.6·10^-3 [6]	—	—
Nb97m	60 sec	—	6.2 [2]	—	—	—
Nb97	72.1 min	—	6.2 [2]	—	—	—
Nb98	52 min	0.20 [6]	0.064 [6]	0.20 [6]	—	—
Mo85	Stable	6.10 [1] 6.11 [6]	6.27 [1,6]	5.03 [6]	—	—
Mo97	"	5.35 [1] 5.37 [6]	6.09 [1,6]	5.65 [6]	—	—

Isotope	T½					
Mo^{98}	"	5.18 [1] / 5.15 [6]	5.78 [1,6]	5.89 [6]	—	—
Mo^{99}	66.0 h	4.8 [1,6]	6.06 [1,6] / 6.1 [2] / 6.14±0.16 [93, 94]	5.9 [1] / 6.10 [6]	6.85 ± 0.41	—
Mo^{100}	≥3·10^{7} yr	4.40 [1] / 4.41 [6]	6.30 [1,6]	7.10 [6]	—	—
Mo^{101}	14.61 min	—	~5.6 [1] / 5.0 [2]	—	—	—
Mo^{102}	11.5 min	—	~4.3 [1] / 4.2 [2]	—	—	—
Mo^{105}	<2 min	—	(0.6) [2]	—	—	—
Tc^{99m}	6.04 h	—	~0.6 [2]	—	—	—
Tc^{99}	2.12·10^{5} yr	—	6.1 [2]	—	—	—
Tc^{101}	14.0 min	—	5.0 [2]	—	—	—
Tc^{102}	5 sec	—	4.2 [2]	—	—	—
Tc^{105}	10 min	—	(0.9) [2]	—	—	—
Tc^{107}	<1.5 min	—	(0.16) [2]	—	—	—
Ru^{101}	Stable	3.00 [1] / 2.91 [6]	5.0 [1,6]	5.91 [6]	—	—
Ru^{102}	"	2.37 [1] / 2.22 [6]	4.1 [6]	5.99 [6]	—	—
Ru^{103}	39.8 days	1.6 [1] / 1.8 [6] / 2.02±0.08 [90]	3.0 [1,6] / 2.9 [2]	5.8 [1] / 5.67 [6]	—	—
Ru^{104}	Stable	0.96 [1] / 0.94 [6]	1.8 [1,6]	5.93 [6]	—	—

TABLE 6 (Continued)

Nuclide	Half-life	U233 [1, 3, 4, 6, 9, 13—15, 20, 26, 88*—90, 97, 104]	U235 [1—3, 4, 6, 14, 20, 38, 51, 80, 88*, 91*, 92*—97*, 104]	Pu239 [1, 3, 4, 6, 14, 20, 87†, 88*, 97*, 98, 99, 100—102, 104]	Am241 ‡ [103]	Cm242 [88*]
Ru105	4.5 h	—	0.9 [1, 2, 6]	—	—	—
Ru106	1.00 yr	0.28 [1] 0.24 [6] 0.259±0.030 [90]	0.38 [1, 2, 6]	5.0 [1] 4.57 [6]	—	—
Ru107	4 min	—	0.2 [2]	—	—	—
Rh102	210 days	—	<5·10^-7 [3]	—	—	—
Rh103m	57 min	—	2.9 [2]	—	—	—
Rh105m	45 sec	—	0.9 [2]	—	—	—
Rh105	36.5 h	0.146±0.037 [90]	0.9 [2]	3.9 [1,6]	—	—
Rh106	30 sec	—	0.38 [2]	—	—	—
Rh107	24 min	—	0.19 [1,6] 0.2 [2]	—	—	—
Rh109	<1 h	—	(0.028) [2]	—	—	—
Pd107	~7·10^6 yr	—	0.2 [2]	—	—	—
Pd109	13.5 h	0.040 [1] 0.044 [6]	0.030 [1,6] 0.028 [2]	1.40 [6] 1.5 [1]	—	—
Pd111	22 min	—	0.018 [2]	—	—	—
Pd112	21 h	0.016 [1,6] 0.0125±0.0004 [90]	0.010 [1,6] 0.011 [2]	0.10 [1] 0.12 [6]	—	—
Ag109m	39.2 sec	—	0.028 [2]	—	—	—

Ag[110m]	253 days	—	$2 \cdot 10^{-7}$ [2]	↑ —	—	—
Ag[110]	24.2 sec	—	$6 \cdot 10^{-9}$ [2]	—	—	—
Ag[111]	7.6 days	0.025 [1] 0.0187±0.0002 [90] 0.024 [6]	0.019 [1,6] 0.018 [2,51]	0.27 [1] 0.23 [6]	0.89±0.05	—
Ag[112]	3.20 h	—	0.011 [2]	—	—	—
Ag[113]	5.3 h	—	0.01 [2]	↓	0.18±0.01	—
Ag[114]	2 min	—	0.01 [2]	—	—	—
Ag[115]	21.1 min	—	0.0077 [1] (∼0.01) [2]	—	—	—
Cd[115m]	43 days	0.001 [1] 0.0011 [6]	0.0007 [1,6] 0.00071 [2]	0.003 [1] 0.0031 [6]	0.004	—
Cd[115]	53 h	0.019 [1] 0.020 [6]	0.0097 [1,6] 0.011 [51] 0.0098 [2]	0.038 [1,6]	0.046± ±0.003	—
Cd[115] (Total)	—	0.020 [1] 0.021 [6]	0.0104 [1,6] 0.0105 [2]	0.041 [1,6]	0.050± ±0.004	—
Cd[117m]	3.0 h	—	0.010 [2] 0.011 [6]	—	—	—
Cd[117]	∼50 min	—	0.010 [2]	—	—	—
Cd[118]	50 min	—	0.01 [2]	—	—	—
In[115m]	4.50 h	—	0.0098 [2]	—	—	—
In[115]	$6 \cdot 10^{14}$ yr	—	0.0099 [2]	—	—	—
In[117m]	1.90 h	—	0.010 [2]	—	—	—
In[117]	1.1 h	—	$2 \cdot 10^{-3}$ [2]	—	—	—

TABLE 6 (Continued)

Nuclide	Half-life	U²³⁸ [1, 3, 4, 6, 9, 13—15, 20, 26, 88*—90, 97, 104]	U²³⁵ [1—3, 4, 6, 14, 20, 38, 51, 80, 88*, 91*, 92*—97*, 104]	Pu²³⁹ [1, 3, 4, 6, 14, 20, 87†, 88*, 97*, 98, 99, 100—102, 104]	Am²⁴¹‡ [103]	Cm²⁴² [88*]
In¹¹⁸	4.5 min	—	0.01 [2]	—	—	—
In¹¹⁹	17.5 min	—	0.01 [2]	—	—	—
Sn¹¹⁷m	14 days	—	$<2 \cdot 10^{-5}$ [2]	—	—	—
Sn¹¹⁹m	~250 days	—	<0.01 [2]	—	—	—
Sn¹²¹	27.5 h	0.018 [1,6]	0.015 [1,6] 0.014 [2]	0.044 [1] 0.043 [6]	0.045± ±0,006	—
Sn¹²³	136 days	$2.5 \cdot 10^{-3}$ [3]	0.0013 [1,6] 0.0012 [2]	—	—	—
Sn¹²³	39.5 min	—	0.014 [2]	—	—	—
Sn¹²⁵	9.4 days	0.050 [1] 0.052 [6]	0.013 [1,6] 0.012 [2]	0.072 [1] 0.071 [6]	—	—
Sn¹²⁵	9.5 min	—	0.011 [2]	—	—	—
Sn¹²⁶	~50 min	—	0.1 [2]	—	—	—
Sn¹²⁷	2.1 h	—	(0.24) [2]	—	—	—
Sn¹²⁸	57 min	—	0.37 [1,6]	—	—	—
Sn¹³⁰	2.6 min	—	2.0 [1.6]	—	—	—
Sb¹²⁵	2.0 yr	—	0.021 [1,6] 0.023 [2]	—	—	—
Sb¹²⁶	9 h	—	0.05 [1] 0.10 [2]	—	—	—

	Half-life					
Sb127	88 h	0.092 [3] 0.59±0.08 [6] 0.60 [6] [90]	0.13 [1,6] 0.25 [2] 0.10 [51]	0.39 [1,6]	—	—
Sb128	10.3 min	—	0.5 [2]		—	—
Sb129	4.2 h	—	1.0 [2]		—	—
Sb130	7.1 min	—	2.0 [2]		—	—
Sb131	23.1 min	—	2.6 [1] (2.7) [2]		—	—
Sb132	2.1 min	—	(3.4) [2]		—	—
Sb133	4.4 min	—	4.0 [1] (3.8) [2]		—	—
Sb134	~50 sec	—	(3.0) [2]		—	—
Te126m	58 days	—	0.003 [2]		—	—
Te127m	105 days	0.067 [3]	0.056 [2] 0.035 [1,6]		—	—
Te127	9.4 h	—	0.25 [2]		—	—
Te129m	33.5 days	0.22 [3]	0.35 [1,6] 0.34 [2]		—	—
Te129	72 min	—	1.0 [2]		—	—
Te130	~10²¹ yr	—	2.0 [2]		—	—
Te131m	30 h	—	0.44 [1,6] 0.44 [2]		—	—
Te131	24.8 min	—	2.9 [2]		—	—
Te132	77.7 h	4.9 [3] 4.4 [6] 4.32±0.25 [90]	~4.7 [1,6] 4.4 [2]	5.2 [1] 5.1 [6]	4.48± ±0.31	—

TABLE 6 (Continued)

Nuclide	Half-life	U²³³ [1, 3, 4, 6, 9, 13—15, 20, 26, 88*—90, 97, 104]	U²³⁵ [1—3, 4, 6, 14, 20, 38, 51, 80, 88*, 91*, 92*—97*, 104]	Pu²³⁹ [1, 3, 4, 6, 14, 20, 87†, 88* 97*, 98, 99, 100—102, 104]	Am²⁴¹ ‡ [103]	Cm²⁴² [88*]
Te¹³³ᵐ	63 min	—	4.9 [1] / 4.6 [2]		—	—
Te¹³³	2 min	—	6.0 [2]		—	—
Te¹³⁴	44 min	—	6.9 [1] / 6.7 [2]		—	—
Te¹³⁵	<2 min	—	(4.2) [2]	—	—	—
I¹²⁸	24.99 min	1.9·10⁻⁴ [88]	3.6·10⁻⁵ [88] / 3·10⁻⁵ [6]	2.0·10⁻⁴ [88]	—	—
I¹²⁹	1.72·10⁷ yr	—	0.9 [1] / 1.0 [2] / 0.8 [6]	—	—	—
I¹³⁰	12.6 h	4.3·10⁻³ [88]	5·10⁻⁴ [6] / 4.5·10⁻⁴ [88]	5.2·10⁻³ [88]	—	—
I¹³¹	8.08 days	2.7 [1] / 2.9 [6] / 2.84±0.20 [90]	~3.1 [1,6] / 2.9 [2]	3.8 [1] / 3.77 [6]	3.7 ± ±0.19	—
I¹³²	2.26 h	—	4.4 [2] / <4.4·10⁻² [88]		—	—
I¹³³	20.8 h	3.37±0.29 [90]	~6.9 [1,6] / 6.5 [2] / 1.2 [3]	5.3 [1] / 5.2 [6]	—	—

Isotope	Half-life							
I[134]	52.5 min	—		—	7.8 [1,6] 1.0 [3] 7.6 [2] 1.0 [88]	—	—	—
I[135]	6.68 h		5.1 [1] 5.5 [6]		6.1 [1,6] 5.9 [2]	5.8 [1] 5.7 [6]	—	—
I[136]	86 sec		1.7 [1] 1.8 [6]		3.1 [1,6] 3.1 [2]	2.1 [1,6]	—	—
I[137]	22.0 sec	—		—	4.9 [2]	—	—	—
I[138]	5.9 sec	—		—	(3.4) [2]	—	—	—
I[139]	2.7 sec	—		—	(1.8) [2]	—	—	—
Xe[128]	Stable	—			$<4\cdot10^{-4}$ [3]	$(2.1\pm0.1)\cdot10^{-4}$ [99]	—	—
Xe[129]	"	—			—	—	—	—
Xe[130]	"	—				$(5.5\pm0.4)\cdot10^{-3}$ [99]	—	—
Xe[131m]	12 days	—		—	0.03 [2]	—	—	—
Xe[131]	Stable		3.74 [1] 3.39 [6]	—	2.93 [1,6] 3.28 [80]	2.87 [1] 3.79±0.11 [99] 3.78 [6] 3.77 [100] 2.71 [101]	—	—
Xe[132]	"		5.10 [1] 4.64 [6]		4.38 [1,6] 4.92 [80]	4.02 [1] 5.26 [100] 5.29±0.16 [99] 5.28 [6] 3.79 [101]	—	—
Xe[133]	5.27 days	—			6.62 [1,6] 6.5 [2]	5.27 [1] 6.95±0.21 [99] 6.91 [6]	—	—
Xe[133m]	2.3 days	—			0.16 [2]	—	—	—

TABLE 6 (Continued)

Nuclide	Half-life	U²³³ [1, 3, 4, 6, 9, 13—15, 20, 26, 88*—90, 97, 104]	U²³⁵ [1—3, 4, 6, 14, 20, 38, 51, 80, 88*, 91*, 92*—97*, 104]	Pu²³⁹ [1, 3, 4, 6, 14, 20, 87†, 88*, 97*, 98, 99, 100—102, 104]	Am²⁴¹ ‡ [103]	Cm²⁴² [88*]
Xe¹³⁴	Stable	6.54 [1] 5.95 [6]	8.06 [1.6] 8.64 [80]	5.69 [1] 7.46 [100] 7.48±0.22 [99] 7.47 [6] 5.37 [101]	—	—
Xe¹³⁵ᵐ	15.6 min	—	1.8 [2]		—	—
Xe¹³⁵	9.13 h	6.0 [20] *⁴	6.41 [20] *⁴ ~0.3 [3] 6.2 [2] ~0.15 [92] 6.3 [6] 0.22 [88]	7.27 [20] *⁴ 7.43 [102]	—	—
Xe¹³⁶	Stable	<8.9 [1] 6.63 [6]	6.46 [1.6] 7.10 [80] 3.4 [88]	5.06 [1] 6.62 [100] 6.70±0.70 [99] 6.63 [6] 4.77 [101]	—	—
Xe¹³⁷	3.9 min		5.9 [2]	—	—	—
Xe¹³⁸	17 min		(5.5) [2]	—	—	—
Xe¹³⁹	41 sec		5.5 [1] (4.7) [2]	—	—	—
Xe¹⁴⁰	16.0 sec	—	3.8 [1] (3.7) [2]	—	—	—

Isotope	Half-life					
Xe[141]	1.7 sec	—	1.34 [1] (1.8) [2]	—	—	—
Xe[143]	1.0 sec	—	0.051 [1] (0.2) [2]	—	—	—
Xe[144]	1 sec	—	(~0) [2]	—	—	—
Cs[133]	Stable	6.18 [1] 5.6 [20] 5.20±0.30 5.50±0.13 } [9, 14] 5.78 [6]	6.59 [6.20] 6.59 [1.95] 7.43 [38]	5.26±0.13 [31] (5.27) [1] 6.90 [100] 6.92 [20] 4.97 [101] 6.91 [6]	—	—
Cs[135]	3.0·10⁶ yr	>4.9 [1] 6.03 [6]	6.41 [1, 6, 95] 6.2 [2] 7.18 [38]	6.95±0,19 [14] 7.25 [100] 5.22 [101] 5.53 [1] 7.17 [6] 5.42 [102]	—	—
Cs[136]	12.9 days	0.12 [6.88] 0.0849±0.0022 [90] 0.118 [104]	6.8·10⁻³ [6] 6.2·10⁻³ [88] 6·10⁻³ [2] 6.85·10⁻³ [104]	1.8·10⁻² [88] 0.09 [3] 0.11 [6] 0.0835 [104]	0.288± ±0.032	0.80
Cs[137]	26.6 yr	7.16 [1] 6.58 [6] 5.80±0.30 } [9.14] 6.16±0.14 5.39 ±0.11 [90]	5.9 [2] 6.81 [38] 6.15 [1. 6. 95]	6.50+0.16 [14] 5.24 [1] 6.48 [100] 6.63 [6] 4.94 [101]	9.20±1.84	—
Cs[138]	32.2 min	—	5.8 [2]	—	—	—
Cs[139]	9.5 min	—	(5.9) [2]	—	—	—
Cs[140]	65 sec	—	(6.0) [2]	—	—	—

4. λ = 0.131 per day is assumed for Xe[135] [20].

TABLE 6 (Continued)

Nuclide	Half-life	U²³³ [1, 3, 4, 6, 9, 13—15, 20, 26, 88*—90, 97, 104]	U²³⁵ [1—3, 4, 6, 14, 20, 38, 51, 80, 88*, 91*, 92*—97*, 104]	Pu²³⁹ [1, 3, 4, 6, 14, 20, 87†, 88*, 97*, 98, 99, 100—102, 104]	Am²⁴¹ ‡ [103]	Cm²⁴² [88*]
Cs¹⁴¹	Short-lived	—	(4.7) [2]	—	—	—
Cs¹⁴²	~1 mln	—	(3.4) [2]	—	—	—
Cs¹⁴³	Short-lived	—	(1.9) [2]	—	—	—
Cs¹⁴⁴	"	—	(1.0) [2]	—	—	—
Ba¹³⁷m	2.60 min	—	5.4 [2]	—	—	—
Ba¹³⁸	>10⁵ yr	—	5.74 [1.6]	6.26±0.15 [14] 6.31 [6, 100] 5.38 [101]	—	—
Ba¹³⁹	84.0 min	6.45 [6]	6.55 [1.6] 6.0 [2]	5.7 [1] 5.87 [6]	6.22±0.31	—
Ba¹⁴⁰	12.8 days	6.0 [1] 5.4 [6] 5.21±0.27 [90]	6.32 (6.44) [1] 6.3 [2] 6.35 [6]	5.68 [1] 5.4 [6]	6.00±0.36	—
Ba¹⁴¹	18 min	—	6.3 [1] 5.9 [2]	—	—	—
Ba¹⁴²	6 min	—	(5.6) [2]	—	—	—
Ba¹⁴³	<0.5 min	—	(4.9) [2]	—	—	—
Ba¹⁴⁴	Short-lived	—	(3.5) [2]	—	—	—

Isotope	Half-life					
La^{139}	Stable	5.91±0.23 [13, 14]		0.747 [87] 6.61 [101]	—	—
La^{140}	40.22 h	2.4·10^{-2} [88]	<0.2 [3] 6.32 (6.44) [1] 6.3 [2] 4.5·10^{-3} [88]	—	—	—
La^{141}	3.8 h	7.1 [6]	6.4 [1,6] 6.0 [2] 0.1 [88]	5.7 [6]	—	—
La^{142}	77 min	—	5.9 [2]	—	—	—
La^{143}	~19 min	—	6.2 [2]	—	—	—
La^{144}	Short-lived	—	(5.8) [2]	—	—	—
Ce^{140}	Stable	5.6 [1] 6.47 [6] 5.45±0.50 [9, 13, 14] 6.16±0.24	6.44 [1, 6] 6.30±0.30 [14]	(5.68) [1] 5.60 [6] 5.52±0.14 [14] 5.58 [100] 1.00 [87] 7.36 [101]	—	—
Ce^{141}	33.1 days	5.30±0.10 [90]	~6.0 [1,6] 6.0 [2]	5.2 [1] 6.94 [101] 5.1 [6]	—	—
Ce^{142}	5·10^{15} yr	5.6 [1] 5.50±0.50 [9, 13, 14] 6.06±0.24 6.83 [6]	5.95 [1] 5.80±0.20 [14] 6.01 [6]	6.66±0.17 [14] 6.69 [1] 4.97 [100] 0.750 [87] 5.01 [6] 6.62 [101]	5.04±0.66	—
Ce^{143}	33 h	6.99±0.35 [90]	5.7 [1, 6] 6.2 [2]	5.4 [1] 5.3 [6]	—	—

TABLE 6 (Continued)

Nuclide	Half-life	U^{238} [1, 3, 4, 6, 9, 13—15, 20, 26, 88*—90, 97, 104]	U^{235} [1—3, 4, 6, 14, 20, 38, 51, 80, 88*, 91*, 92*—97*, 104]	Pu^{239} [1, 3, 4, 6, 14, 20, 87†, 88*, 97*, 98, 99, 100—102, 104]	Am^{241} ‡ [103]	Cm^{242} [88*]
Ce^{144}	285 days	4.1 [1] 3.69±0.18 [90] 4.5 [6]	~6.0 [1, 6] 6.1 [2] 6.0 [93]	3.6 [98] 3.79 [6] 5.29 [1] 0.0238 [87]	3.15±0.41	—
Ce^{145}	3.0 min	—	4.2 [2]	—	—	—
Ce^{146}	13.9 min	—	(3.2) [2]	—	—	—
Pr^{141}	>2·10^{16} yr	5.57±0.19 [13, 14] 6.4 [6]	5.60±0.30 [14]	6.02±0.18 [14] 0.700 [87] 4.5 [6]	3.32±0.46	—
Pr^{143}	13.76 days	—	6.2 [2]	—	—	—
Pr^{144}	17.27 min	—	6.1 [2]	—	—	—
Pr^{145}	5.95 h	—	4.2 [2]	—	—	—
Pr^{146}	24.4 min	—	3.3 [2]	—	—	—
Nd^{142}	Stable	5.00±0.30 5.19±0.17 } [9, 13, 14] 5.2 [1] 5.99 [6] 6.45 [20]	—	0.0944 [87]	—	—
Nd^{143}	"		5.80±0.20 [14] 5.98 [1] 5.80 [20] 5.40 [96] 6.03 [6]	6.31 [1] 4.49 [20] 6.10±0.15 [14] 0.00622 [87] 4.56 [100] 4.57 [6] 5.98 [101]	—	—

Isotope	Half-life					
Nd^{114}	$5\cdot10^{15}$ yr	4.0 [1] 4.61 [6] 3.80±0,40 3.84±0,15 }[9, 13, 14]	5.67 [1] 4.64 [96] 6.1 [2] 5.62 [6] 5.60±0.30 [14]	(5.29) [1] 3.93 [6] 5.50±0.17 [14] 3.84 [100] 0.845 [87] 5.00 [101]	—	—
Nd^{145}	Stable	3.0 [1] 3.47 [6] 2.82±0.25 2.88±0.08 }[9, 13, 14]	3.95 [1] 3.98 [6] 4.00±0.10 [14] 3.62 [96]	4.20±0.11 [14] 4.24 [1] 3.13 [6] 0.147 [87] 3.12 [100] 4.07 [101]	—	—
Nd^{146}	Stable	2.3 [1] 2.63 [6] 2.20±0.15 2.24±0.07 }[9, 13, 14]	3.07 [1, 6] 3.20 [14] 2.81 [96]	3.53 [1] 2.60 [6] 3.53±0.09 [14] 2.57 [100] 0.493 [87] 3.36 [101]	—	—
Nd^{147}	11.06 days	—	~2.7 [1, 6] 2.6 [2] 2.6 [33]	2.2 [6, 98]	2.06±0.33	—
Nd^{148}	Stable	1.15 [1] 1.34 [6] 1.03±0.10 1.07±0.04 }[9, 13, 14]	1.70 [1] 1.71 [6] 1.70±0.10 [14] 1.64 [96]	2.30±0.05 [14] 2.28 [1] 1.73 [6] 0.194 [87] 1.71 [100] 2.27 [101]	—	—
Nb^{149}	2.0 h	—	1.3 [2]	—	—	—

TABLE 6 (Continued)

Nuclide	Half-life	U233 [1, 3, 4, 6, 9, 13—15, 20, 26, 88*—90, 97, 104]	U235 [1—3, 4, 6, 14, 20, 38, 51, 80, 88*, 91*, 92*—97*, 104]	Pu239 [1, 3, 4, 6, 14, 20, 87†, 88* 97*, 98, 99, 100—102, 104]	Am241 ‡ [103]	Cm242 [88*]
Nd150	> 10^16 yr	0.48 [1] 0.51±0.08 0.49±0.02 } [9, 13, 14] 0.56 [6]	0.67 [1, 6] 0.74 [2] 0.658 [96] 0.70±0.10 [14]	1.35±0.02 [14] 1.38 [1] 0.102 [87] 1.01 [6] 1.02 [100] 1.31 [101]	—	—
Nd151	15 min	—	(0.48) [2]	—	—	—
Pm147	2 64 yr	1.53±0.06 [13, 14] 2.1 [20] 1.9 [6]	2.6 [2] 2.38 [20] 2.90±0.40 [14]	2.58±0.05 [14] 2.07 [20] 1.94 [6]	—	—
Pm148	5.3 days	—	<2·10^-4 [4]	—	—	—
Pm149	54 h	—	1.3 [2, 93]	1.4 [6, 98]	—	—
Pm151	27.5 h	—	0.5 [2]	—	—	—
Sm147	1.3·10^11 yr	1.71 [1] 1.98 [6]	2.38 [1] 2.36 [6] 2.15 [96] 2.6 [2]	0.0105 [87] 2.92 [1] 1.99 [100] 2.07 [6] 2.81 [101]	—	—
Sm148	Stable	—	—	0.119 [87] 1.30 [100] 1.81 [101]	—	—

Isotope	Half-life					
Sm149	Stable	0.61 [1] 0.76 [6] 0.8 [20] 0.66±0.13 0.70±0.03	1.13 [1, 6, 20] 1.10 [96] 1.50±0,30 [14]	1.68±0.02 [14] 1.89 [1] 1.32 [6. 20]	—	—
Sm150		} [9, 13, 14]	—	0.0438 [87]	—	—
Sm151	~93 yr	0.26 [1] 0.335 [6] 0.33±0.03 [13, 14] 0.3 [20] —	0.45 [1, 20] 0.5 [2] 0.44 [6] 0.445 [96]	1.01±0.02 [14] 0.80 [6] 1.17 [1] 0.79 [20] 0.000477 [87] 0.802 [100] 1.10 [101]	—	—
Sm152	Stable	0.17 [1] 0.21±0,02 [13, 14] 0.22 [6]	0.285 [1] 0.279 [96] 0.281 [6]	0.75±0.015 [14] 0.83 [1] 0.62 [6] 0.0231 [87] 0.616 [100] 0.88 [101]	—	—
Sm153	47.1 h	0.095 [1] 0.11 [6]	0.15 [1, 2, 6] 0.14 [93]	0.33 [98] 0.41 [1] 0.37 [6]	0.76±0.12	—
Sm154	Stable	0.037 [1] 0.045 [6]	0.077 [1, 6] 0.0908 [96]	0.36±0.009 [14] 0.32 [1] 0.29 [6] 0.0415 [87] 0.293 [100] 0.40 [101]	—	—
Sm155	23.5 min	—	0.033 [1, 6] 0.031 [2]	0.22 [1] 0.23 [6]	—	—
Sm156	9.0 h	—	0.013 [1, 2]	—	—	—

TABLE 6 (Continued)

Nuclide	Half-life	U233 [1, 3, 4, 6, 9, 13—15, 20, 26, 88*—90, 97, 104]	U235 [1—3, 4, 6, 14, 20, 38, 51, 80, 88*, 91*, 92*—97*, 104]	Pu239 [1, 3, 4, 6, 14, 20, 87†, 88*, 97*, 98, 99, 100—102, 104]	Am241 ‡ [103]	Cm242 [88*]
Eu153	Stable	0.13±0.02 [6, 13, 14]	0.170 [96] 0.169 [6]	0.0235 [87]	—	—
Eu154	16 yr	—	—	0.00430 [87]	—	—
Eu155	1.7 yr	—	~0.03 [1] 0.033 [6]	0.00254 [87]	—	—
Eu156	15.4 days	0.011 [6]	0.031 [2] 0.014 [1, 6] 0.013 [2, 93]	0.10 [98] 0.12 [1] 0.11 [6]	—	—
Eu157	15.4 h	—	0.0078 [1, 6] 0.0074 [2]	—	—	—
Eu158	60 min	—	0.002 [1, 2, 6]	—	—	—
Gd155	Stable	—	0.0500 [96]	0.30 [101] 0.478 [87] 0.08 [101]	—	—
Gd156	"	—	0.0260 [96]	—	—	—
Gd157	"	—	0.0150 [96] 0.0084 [96]	0.0615 [87] 0.021 [6, 98]	—	—
Gd158	"	—	0.00107 [1, 6] 0.0011 [2, 93]	—	—	—
Gd159	18.0 h	—	0.0027 [96]	—	—	—
Gd160	Stable	—	$8 \cdot 10^{-3}$ [2]	0.00162 [87]	—	—
Gd161	3.6 min	—	$7.6 \cdot 10^{-5}$ [1, 6]	—	—	—
Tb161	6.88 days	—	$8 \cdot 10^{-5}$ [2] $7.8 \cdot 10^{-5}$ [93]	0.0039 [6, 98]	—	—
Dy166	82 h	—	—	0.000068 [6, 98]	—	—

TABLE 7. Averaged, Extrapolated, and Interpolated Yields of Fission Fragments on U^{233}, U^{235}, Pu^{239} Fissioned by Thermal Neutrons (based on all available experimental data) (%)

Mass number	U^{233}		U^{235} [141, 144, 145]	Pu^{239} [141]
	[146]	[141]		
72	—	—	$1.5 \cdot 10^{-5}$ [144]	—
73	—	—	$1.0 \cdot 10^{-4}$ [144]	—
74	—	—	$3.5 \cdot 10^{-4}$ [144]	—
77	—	—	0.008 [144]	—
78	—	—	0.020 [144]	—
79	—	—	0.055 [144]	—
80	0.15	—	0.11 [144]	--
81	0.25	0.45	0.21 [144]; 0.14 [141]	0.024
82	0.50	0.7	0.35 [144]; 0.28 [141]	0.045
83	1.18	1.2	0.548 [144]; 0.544 [141]	0.084
84	1.97	1.9	1.01 [144]; 1.0 [141]	0.20
85	2.54	1.9	1.31 [144]; 1.0 [141]	0.22
86	3.30	3.2	2.04 [144]; 2.02 [141]	0.5
87	4.61	4.0	2.50 [144]; 2.49 [141]	0.7
88	5.54	5.0	3.58 [144]; 3.57 [141]	1.4
89	6.15	6.5	4.73 [144]; 4.79 [141]	1.9
90	6.75	6.5	5.77 [144]; 5.77 [141]	2.5
91	6.45	6.5	5.97 [144]; 5.84 [141]	2.9
92	6.72	6.7	6.03 [144]; 6.03 [141]	3.8
93	7.01	7.0	6.51 [144]; 6.45 [141]	4.5
94	6.68	6.8	6.55 [144]; 6.40 [141]	5.0
95	6.23	6.2	6.55 [144]; 6.27 [141]	5.7
96	5.67	5.7	6.41 [144]; 6.33 [141]	5.9
97	5.51	5.3	6.33 [144]; 6.09 [141]	5.9
98	5.22	5.2	5.93 [144]; 5.78 [141]	6.0
99	4.84	4.8	6.25 [144]; 6.06 [141]	6.1
100	4.49	4.4	6.58 [144]; 6.30 [141]	6.0
101	2.87	3.0	5.0 [144]; 5.0 [141]	6.0
102	2.10	2.4	4.1 [144]; 4.1 [141]	6.0
103	1.56	1.6	2.9 [144]; 3 [141]	5.7
104	0.94	0.97	1.8 [144]; 1.8 [141]	5.1
105	0.43	0.5	0.90 [144]; 0.9 [141]	4.7

TABLE 7 (Continued)

Mass number	U²³³		U²³⁵ [141, 144, 145]	Pu²³⁹ [141]
	[146]	[141]		
106	0.22	0.28	0.38 [144]; 0.38 [141]	4.0
107	−	0.15	0.17 [144]; 0.19 [141]	3.0
108	−	0.06	0.07 [144]; 0.07 [141]	2.0
109	−	0.04	0.030 [144]; 0.03 [141]	1.5
110	−	0.03	0.020 [144]; 0.024 [141]	0.7
111	−	0.025	0.016 [144]; 0.019 [141]	0.27
112	−	0.02	0.013 [144]; 0.01 [141]	0.1
113	−	0.02	0.012 [144]; 0.01 [141]	0.06
114	−	0.02	0.011 [144]; 0.01 [141]	0.05
115	−	0.02	0.011 [144]; 0.01 [141]	0.04
116	−		0.011 [144]	
117	−	−	0.010 [145]	−
118	−	−	0.010 [145]	−
119	−	−	0.011 [145]	−
120	−	−	0.011 [145]	−
121	−	−	0.012 [145]	−
122	−	−	0.013 [145]	−
123	−	−	0.014 [145]	−
124	−	−	0.017 [145]	−
125	−	−	0.036 [145]	−
1.6	−	0.24	0.10 [145]; 0.05 [141]	0.25
127	−	0.39	0.25 [145]; 0.13 [141]	0.39
128	−	1.0	0.50 [145]; 0.37 [141]	0.80
129	−	2.0	1.0 [145]; 0.9 [141]	1.4
130	−	2.7	2.0 [145]; 2.0 [141]	2.0
131	3.52	3.7	2.93 [145]; 2.93 [141]	3.2
132	4.82	5.1	4.38 [145]; 4.38 [141]	4.0
133	5.77	6.2	6.62 [145]; 6.59 [141]	5.4
134	6.18	6.6	8.06 [145]; 8.06 [141]	5.8
135	6.02	6.7	6.45 [145]; 6.41 [141]	5.5
136	6.89	6.9	6.47 [145]; 6.46 [141]	5.1
137	6.58	7.2	6.17 [145]; 6.15 [141]	5.2
138	6.73	6.8	6.68 [145]; 5.74 [141]	5.3
139	6.71	6.4	6.42 [145]; 6.55 [141]	5.8

TABLE 7 (Continued)

Mass number	U²³³ [146]	U²³³ [141]	U²³⁵ [141, 144, 145]	Pu²³⁹ [141]
140	6.72	6.1	6.25 [145]; 6.44 [141]	5.8
141	6.92	5.9	5.73 [145]; 6.4 [141]	6.0
142	7.00	5.7	5.80 [145]; 5.95 [141]	6.8
143	6.22	5.2	5.71 [145]; 5.98 [141]	6.1
144	4.87	4.1	5.30 [145]; 5.67 [141]	5.3
145	3.66	3.0	3.80 [145]; 3.95 [141]	4.1
146	2.74	2.3	2.89 [145]; 3.07 [141]	3.6
147	2.08	1.7	2.16 [145]; 2.38 [141]	2.6
148	1.40	1.2	1.61 [145]; 1.7 [141]	2.3
149	0.790	0.62	1.02 [145]; 1.13 [141]	1.7
150	0.567	0.48	0.628 [145]; 0.67 [141]	1.4
151	0.334	0.26	0.399 [145]; 0.45 [141]	1.0
152	0.222	0.17	0.260 [145]; 0.285 [141]	0.75
153	0.122	0.095	0.148 [145]; 0.15 [141]	0.43
154	0.048	0.037	0.0724 [145]; 0.077 [141]	0.32
155	0.033	0.015	0.0291 [145]; 0.03 [141]	0.21
156	—	0.005	0.015 [145]; 0.013 [141]	0.12
157	—	0.0025	0.007 [145]; 0.0078 [141]	0.07
158	—	0.001	0.002 [145]; 0.002 [141]	0.03
159	—	0.0005	0.001 [145]; 0.001 [141]	0.015

Remark: The yields were determined experimentally in [144-146] only for several mass numbers.

TABLE 8. Yields of Fragments in Fission of U²³⁵ by Resonance Neutrons [51]

Nuclide	Half-life	Fragment yields (%) in fission by resonance neutrons of various energies (eV)			
		1.1	3.1	9.5	Thermal
Sr⁸⁹	50.5 days	4.8	4.8	4.8	4.8
Ag¹¹¹	7.6 days	0.020	0.019	0.018	0.018
Cd¹¹⁵	53 h	0.013	0.008	0.010	0.011
Sb¹²⁷	88 h	0.11	—	—	0.10

TABLE 9. Relative Probability of Asymmetrical Fission of U^{233} by Resonance Neutrons [50]

Neutron energy, eV	$\dfrac{\left[\dfrac{Y(Mo^{99})}{Y(Ag^{111})}\right]_{res}}{\left[\dfrac{Y(Mo^{99})}{Y(Ag^{111})}\right]_{th}}$	$\dfrac{\left[\dfrac{Y(Mo^{99})}{Y(Cd^{115})}\right]_{res}}{\left[\dfrac{Y(Mo^{99})}{Y(Cd^{115})}\right]_{th}}$
1.8	1.217 ± 0.038	—
2.3	1.200 ± 0.059	1.415 ± 0.103
4.7	0.957 ± 0.022	1.031 ± 0.168
All epicadmium neutrons	1.088 ± 0.023	1.200 ± 0.052

TABLE 10. Relative Probability of Asymmetrical Fission of Pu^{239} and Pu^{241} by Resonance Neutrons [53]

Nuclides	Neutron energy, eV	$\dfrac{\left[\dfrac{Y(Mo^{99})}{Y(Cd^{115})}\right]_{res}}{\left[\dfrac{Y(Mo^{99})}{Y(Cd^{115})}\right]_{th}}$	$\dfrac{\left[\dfrac{Y(Mo^{99})}{Y(Sn^{121})}\right]_{res}}{\left[\dfrac{Y(Mo^{99})}{Y(Sn^{121})}\right]_{th}}$	$\dfrac{\left[\dfrac{Y(Mo^{99})}{Y(Sn^{125})}\right]_{res}}{\left[\dfrac{Y(Mo^{99})}{Y(Sn^{125})}\right]_{th}}$
Pu^{239}	0.06	1.33 ± 0.06	1.26 ± 0.05	1.16 ± 0.04
	0.22	2.60 ± 0.15	2.82 ± 0.32	1.85 ± 0.18
	0.297	3.00 ± 0.28	3.28 ± 0.31	2.05 ± 0.07
	0.36	3.24 ± 0.14	3.22 ± 0.25	1.78 ± 0.16
	All epicadmium neutrons	2.41 ± 0.15	2.34 ± 0.06	1.79 ± 0.06
Pu^{241}	Same	1.02 ± 0.03	1.01 ± 0.08	0.995 ± 0.15

TABLE 11. Relative Yield of Products of Am^{241} Fission by Thermal and Resonance Neutrons [54]

Nuclide	Half-life	Thermal neutrons	First resonance (0.3 eV)
Mo^{99}	66.0 h	1.0	1.0
Ag^{111}	7.6 days	0.126 ± 0.02	0.125 ± 0.034
Cd^{115}	53 h	$(6.9\pm1.0)\cdot10^{-3}$	$(8.0\pm2.3)\cdot10^{-3}$
Sn^{121}	27.5 h	$(2.0\pm0.6)\cdot10^{-3}$	$(2.1\pm1.3)\cdot10^{-3}$
Sn^{125}	9.4 days	$(5.3\pm1.4)\cdot10^{-3}$	$(4.1\pm2.9)\cdot10^{-3}$
Sb^{127}	88 h	0.074 ± 0.004	0.071 ± 0.010
Ba^{110}	12.8 days	0.92 ± 0.04	0.85 ± 0.08

TABLE 12. Fragment Yields in Fission of Th^{232}, U^{235}, and U^{238} by Neutrons of about 8 MeV Energy (%)

Nuclide	Half-life	Th^{232} [1.112]	U^{235} [93]	U^{238} [93]
Ge^{77}	11.3·h	0.022 [1. 112]	—	—
As^{77}	38.7 h	0.052 [112]	—	—
Br^{83}	2.30 h	2.74 [112]	—	—
Kr^{83}	Stable	2.7 [1]	—	—
Sr^{89}	50.5 days	(6.7±0.7) * [1, 112]	—	.—
Sr^{91}	9.67 h	5.6 [1, 112]	—	---
Zr^{97}	17.0 h	4.95 [112] 5.0 [1]	—	—
Mo^{99}	66.0 h	3.1 [1, 112]	5.4	6.2
Ru^{103}	39.8 days	0.51 [112] 0.5 [1]	—	—
Ru^{106}	1.00 yr	0.53 [1, 112]	—	—
Ag^{111}	7.6 days	0.63 [1, 112]	—	---
Cd^{115}	53 h	0.76 [1, 112]	---	—
Cd^{117m}	3.0 h	0.37 [112]	—	—
Te^{132}	77.7 h	1.8 [112]	—	—
I^{131}	8.08 days	2.3 [112]	—	—
Ba^{139}	84 min	9.0 [112]	—	—
Ce^{144}	285 days	7.2 [112]	3.2 4.0	3.9 4.3
Nd^{147}	11.06 days	—	1.9 2.2	2.7
Pm^{149}	54 h	—	1.3 1.2	1.9
Sm^{153}	47.1 h	—	0.18 0.19	0.41
Eu^{154}	15.4 days	—	0.035	0.092 0.087
Gd^{159}	18.0 h	—	0.0068 0.0058	0.018 0.016
Tb^{161}	6.88 days	—	0.0019 0.0020	0.0044 0.0041

*The yield assigned to Sr^{89} is the same as in the fission of Th^{232} by fission spectrum neutrons (cf. Table 13).

TABLE 13. Fragment Yields in the Fission of Th^{232}, U^{233}, U^{235}, U^{238}, and Pu^{239} by Fission Spectrum Neutrons* (%)

Nuclide	Half-life	Th^{232} [1, 4, 6, 105, 106]	U^{233} [31]	U^{235} [6, 44, 93, 107, 109, 110]	U^{238} [1, 3, 4, 6, 44, 80, 93, 108, 110, 111]	Pu^{239} [1, 6, 31, 107, 109, 110]
Zn^{72}	49 h	0.00033±0.00008 [1, 6, 105]	—	—	—	—
Ga^{73}	5.0 h	0.00045±0.00022 [1, 6, 105]	—	—	—	—
Ge^{77}	11.3 h	0.009±0.002 [1, 6, 105]	—	—	4.3·10^{-3} [3]	—
As^{77}	38.7 h	0.020±0.007 [1, 6, 105]	—	—	0.0038 [1, 6]	—
Br^{83}	2.3 h	1.9±0.45 [1, 6, 105]	—	—	—	—
Kr^{83}	Stable	1.99±0.01 [1, 6, 106]	—	—	0.40'[1, 6] / 0.47 [80]	—
Kr^{84}	"	3.65±0.02 [1, 6, 106]	—	—	0.85 [1, 6] / 0.98 [80]	—
Kr^{85}	10.3 yr	0.87 [1, 6] / 3.88±0.02 [106]	—	—	0.153 [1, 6]	—
Kr^{86}	Stable	6.00±0.03 [1, 6, 106]	—	—	1.38 [1, 6] / 1.63 [80]	—
Sr^{89}	50.5 days	6.7±0.7 [1, 6, 105]	6.30±0.60 [31]	4.4±0.4 [44] / 5.6±0.4 [110]	2.9 [1.6] / 4.4±0.4 [44] / 3.7±0.3 [110]	1.8±0.2 [110]

Isotope	Half-life			5.0 [6]	3.2 [1, 6]	2.2 [1, 6]
Sr⁹⁰	27.7 yr	6.1±1.2 [105] 6.8 [1.6]	—	5.0 [6]	3.2 [1, 6]	2.2 [1, 6]
Sr⁹¹	9.67 h	6.4±0.7 [105] 7.2 [1.6]	—	—	—	—
Zr⁹⁵	65 days	—	—	5.85±0.55 [44] 7.7±0.6 [110]	5.0±0.5 [44] 5.7 [1, 6] 6.5±0.6 [110]	5.6 [3] 5.3±0.5 [110]
Zr⁹⁷	17.0 h	5.2 [1.6] 5.4±0.8 [105]	—	6.55±0.70 [44]	5.2±0.6 [44]	5.2 [1, 6]
Mo⁹⁹	66.0 h	2.7 [1.6[2.9±0.3 [105]	4.75±0.35 [31]	6.1 [6, 93] 5.9±0.4 [44] 6.4±0.4 [109, 110]	7.0±0.7 [44] 6.3 [1, 6] 7.0±0.8 [108] 6.2 }[93] 6.2 6.6±0.4 [110]	5.9 [1] 5.5±0.4 [109] 5.9±0.6 [31] 6.0 [6]
Ru¹⁰³	39.8 days	0.16 [1. 6] 0.20±0.07 [105]	0.413±0.045 [31]	3.75±0.55 [44] 3.2±0.6 [109]	3.9±0.5 [44] 6.6 [1, 6]	6.0±0.7 [31] 5.7±1.0 [109]
Ru¹⁰⁶	1.00 yr	0.042 [1, 6] 0.058±0.006 [105]	0.16±0.02 [31]	1.19±0.14 [44] 0.71±0.12 [109]	2.85±0.30 [44] 2.7 [1, 6]	4.8±0.6 [31] 4.6±0.8 [109]
Rh¹⁰⁵	36.5 h	0.07±0.02 [1, 6, 105]	—	1.45±0.15 [44]	3.5±0.4 [44]	—
Pd¹⁰⁹	13.5 h	0.055 [1, 6] 0.053±0.010 [105]	—	0.146 [6]	0.13 [44] 0.32 [1, 6]	1.9 [1] 2.0 [6]
Pd¹¹²	21 h	0.057 [1, 6] 0.065±0.010 [105]	—	0.041 [6]	0.07 [44] 0.046 [1, 6]	0.14 [1, 6]
Ag¹¹¹	7.6 days	0.052±0.010 [1, 6, 105]	0.0837±0.008 [31]	0.035±0.007 [44] 0.031±0.002 [110] 0.071 [6]	0.094±0.012 [44] 0.073 [1] 0.094±0.008 [110] 0.076 [6]	0.55±0.06 [31] 0.45±0.03 [110]

• In several papers, the irradiations reported were by fission neutrons and by neutrons of spectra close to the fission spectrum.

TABLE 13 (Continued)

Nuclide	Half-life	Th232 [1, 4, 6, 105, 106]	U233 [31]	U235 [6, 44, 93, 107, 109, 110]	U238 [1, 3, 4, 6, 44, 80, 93, 108, 110, 111]	Pu239 [1, 6, 31, 107, 109, 110]
Cd115m	43 days	0.003±0.0015 [1, 6, 105]	—	0.0022 [44]	0.003 [1, 6, 44]	0.069 [1]
Cd115	53 h	0.072±0.014 [1, 6, 105]	0.052±0.006 [31]	0.0304±0.006 [44] / 0.022±0.002 [110] / 0.038 [6]	0.034 [6] / 0.046±0.007 [44] / 0.037 [1] / 0.042±0.004 [110]	0.09±0.01 [31] / 0.098±0.008 [1,10] / 0.067 [6]
Cd115 (Total)	—	0.075±0.015 [1, 6, 105]	0.056±0.006 [31]	0.0326 [44]	0.040 [1] / 0.049±0.007 [44]	0.095±0.010 [31]
Sn125	9.4 days	—	—	—	0.037 [6]	—
Sb127	88 h	—	—	—	0.078±0.012 [44]	—
Te129m	33.5 days	—	0.602±0.050 [31]	0.55±0.06 [44]	0.12 [1,6] / 0.17±0.02 [44]	0.45±0.09 [31]
Te129 (Total)	—	—	1.57 [31]	—	0.26±0.03 [44]	1.17 [31]
Te132	77.7 h	2.4±0.7 [1, 6, 105]	4.36±0.40 [31]	5.35±0.50 [44]	4.7 [1, 6] / 4.1±0.4 [44]	3.5±1.0 [31]
I131	8.08 days	1.2±0.6 [1, 6, 105]	—	—	—	—
Xe131	Stable	1.62±0.01 [1, 6, 106]	—	—	3.2 [1, 6]	—
Xe132	"	2.87±0.02 [1, 6, 106]	—	—	4.7 [1, 6]	—
Xe134	"	5.38±0.03 [1, 6, 106]	—	—	6.6 [1, 6]	—
Xe136	"	5.65±0.03 [1, 6, 106]	—	—	5.9 [1, 6]	—

Isotope	Half-life					
Cs133	Stable	—	—	—	5.5 [1, 6]	—
Cs135	3.0·10^6 yr	—	—	—	6.0 [1, 6]	—
Cs136	12.9 days	0.0017+0.0009−0.0017 [105]	0.11 [31]	—	0.035±0.007 [44]	—
Cs137	26.6 yr	<1.7·10^-3 [4] 6.3 [1, 6] 6.6±1.0 [105]	6.28±0.50 [31]	6.87±0.17 [107] 6.3 [6]	6.1±0.7 [44] 6.2 [1, 6]	6.6 [1] 7.45±0.20 [107] 6.8 [6]
Ba139	84.0 min	—	6.31±0.50 [31]	5.0±0.4 [44]	5.1 [3] 5.8±0.5 [44]	5.4±0.5 [31]
Ba140	12.8 days	6.2±2.0 [1, 6, 105]		6.0±0.5 [110]	5.7 [1, 6] 6.7±0.5 [110]	5.0 [1, 6] 4.9±0.4 [110]
Ce141	33.1 days	9.0±3.0 [1, 6, 105]	6.74±0.60 [31]	6.1±0.6 [44]	4.0 4.4 }[93] 4.5 [6] 4.9 [1]	—
Ce144	285 days	7.1±1.0 [1, 6, 105]	—	5.3 5.1 }[93] 5.0 [6]		
Nd147	11.06 days	—	—	2.5 2.4 }[93] 2.3 [6]	2.6 2.9 }[93] 2.6 [6] 2.2	—
Pm149	54 h	—	—	1.3 1.3 }[93] 1.1 [6]	2.0 1.8 [6]	—
Sm153	47.1 h	—	—	0.18 0.19 }[93] 0.21 [6]	0.39 }[93] 0.37 0.41 [6]	0.48 [1, 6]
Eu156	15.4 days	—	—	0.023 0.023 }[93] 0.025 [6]	0.066 [1] 0.073 }[93] 0.066 0.071 [6]	—
Gd159	18.0 h	-	—	0.0030 0.0038 }[93] 0.0034 [6]	0.0095 }[93] 0.0077 0.0084 [6]	—
Tb161	6.88 days	—	—	0.00045 0.00046 }[93] 0.00046 [6]	0.0016 }[93] 0.0016 0.0016 [6]	—

63

TABLE 14. Fragment Yields in Fission of Th²³², U²³³, U²³⁵, U²³⁸, Np²³⁷, Pu²³⁹ by Neutrons of 14.5 MeV Energy (%)

Nuclide	Half-life	Th²³² [21]	U²³³ [31]	U²³⁵ [1, 6, 44, 88*; 113*—115]	U²³⁸ [6, 44, 94, 108, 113†, 115, 116, 152]	Np²³⁷ [117]	Pu²³⁹ [31, 113†]
Ga⁷³	5.0 h	<0.06	—	—	—	—	—
Br⁸²	35.9 h	—	—	0.004 [6]		—	—
Br⁸³	2.30 h	1.6±0.3	—	1.16 [1, 6]	0.62 [6]	—	—
Br⁸⁴	6.0 min	—	—	—	1.1 [6]	—	—
Sr⁸⁹	50.5 days	5.7±0.8	—	4.5 [1, 6] 4.2±0.4 [44] 0.86±0.04 [113]	3.3±0.3 [44] 2.30±0.12 [116] 0.55±0.03 [113] 2.7 [6] 2.0±0.2 } [152] 3.0	—	0.44±0.02 [113]
Sr⁹⁰	27.7 yr	—	—	4.5 [1, 6]	3.1 [6] 3.4±0.3 [152]	—	—
Sr⁹¹	9.67 h	—	—	4.9 [1, 6] 0.96±0.07 [113]	0.65±0.05 [113] 3.6 [6] 2.6±0.3 [152]	2.71±0.25	0.49±0.03 [113]
Y⁹¹	57.5 days	5.2±0.8	—	—	2.78±0.14 [116] 2.8 [6]	—	—
Y⁹³	10.4 h	—	—	—	4.5 [6]	4.94±0.25	—
Zr⁹⁵	65 days	6.7±1.5	—	4.3±0.4 [44] 0.97±0.04 [113] 5.0 [6]	4.6±0.4 [44] 0.93±0.04 [113] 5.2 [6]	—	—

Isotope	Half-life						
Zr97	17 h	—	—	5.6 [6] 4.4±0.4 [44] 5.4 [1] 1.16±0.05 [113]	4.9±0.4 [44] 1.02±0.05 [113] 5.8 [6] 4.8 [152]	5.43±0.49	0.96±0.04 [113]
Mo99	66.0 h	2.0±0.2	3.5±0.3	5.65±0.4 [44] 5.17 [1, 6] 1 [113] 5.01±0.15 [115]	6.5±0.5 [44] 5.68±0.14 [94] 5.58±0.28 [116] 1 [113] 5.7 [6]	4.94	— [113] 4.16±0.40 [31]
Mo101	14.61 min	—	—	—	5.86±0.16 [115] 6.5±0.7 [108] 5.7 [152] 0.99±0.04 [113] 5.5 [6]	—	—
Mo102	11.5 min	—	—	—	0.71±0.08 [113] 3.9 [6]	—	—
Ru103	39.8 days	—	2.31±0.30	3.25±0.3 [44] 3.5 [1, 6] 0.28±0.02 [113]	3.0±0.3 [44]	—	6.25±0.80 [31]
Ru105	4.5 h	—	—	—	0.39±0.03 [113] 2.3 [6]	—	—
Ru106	1.00 yr	—	1.52±0.20	2.3±0.2 [44] 1.58 [1, 6]	2.4±0.3 [44]	—	4.16±0.5 [31]
Rh105	36.5 h	—	—	2.95±0.3 [44] 1.85 [1] 1.70 [6]	3.3±0.3 [44] 3.4 [6] 3.4 [152]	3.50±0.20	—
Pd109 Pd112	13.5 h 21 h	— —	— —	1.31 [1, 6] 0.81 [1, 6]	1.2 [6] 0.7 [44] 0.69 [6]	1.48±0.25 1.23±0.05	— —
Ag111	7.6 days	1.27±0.15	1.22±0.12	1.05±0.10 [44] 1.24 [1] 0.22±0.01 [113] 1.20 [6]	1.06±0.12 [44] 0.81±0.04 [116] 0.18±0.01 [113] 0.96 [6] 0.6±0.1 } [152] 0.87	1.23±0.05	1.46±0.14 [31] 0.34±0.02 [113]

* Only independent yields listed in [88].
† Only relative yields listed in [113].
‡ Estimate of total chain yield.

65

TABLE 14 (Continued)

Nuclide	Half-life	Th232 [21]	U233 [31]	U235 [1, 6, 44, 88*, 113*—115]	U238 [6, 44, 94, 108, 113†, 115, 116, 152]	Np237 [117]	Pu239 [31, 113†]
Ag113	5.3 h	—	—	0.22±0.02 [113] 1.1 [6]	0.16±0.01 [113] 0.85 [6] 0.6±0.1 [152]	—	—
Cd115m	43 days	—	—	0.06 [44] 0.062 [1,6] 0.069 [114]	0.06 [6.44] 0.06±0.01 [116]	—	—
Cd115	53 h	—	0.98±0.18	0.98 [114] 0.95±0.10 [44] 0.88 [1] 0.21±0.01 [113] 1.00 [6]	0.58±0.03 [116] 0.80±0.09 [44] 0.16±0.01 [113] 0.64 [6] 0.71 [152]	—	0.28±0.02 [113] 1.23±0.10 [31]
Cd115 (Total)	—	1.07±0.12	1.05±0.20	1.0±0.1 [44] 0.94 [1] 1.06 [6]	0.86±0.10 [44] 0.70 [6]	1.23±0.05	1.03±0.11 [31]
Sn121	27.5 h	—	—	1.1±0.1 [44] 1.23 [1,6]	0.96 [6] 0.73 [152]	—	—
Sn125	9.4 days	—	—	1.34 [1,6]	0.45 [6] 0.83 [152]	—	—
Sb127	88 h	—	—	2.28 [1,6]	1.7 [6] 1.43 [152]	2.52±0.15	—
Sb129	4.2 h	—	—	—	1.4 [6]	—	—

Isotope	Half-life						
Te¹²⁹ᵐ	33.5 days	0.73±0.18	—	1.58±0.12 [44]	1.22±0.09 [44]	—	—
Te¹³¹ᵐ	30 h	—	—	1.6 [88]	—	—	—
Te¹³²	77.7 h	2.8±0.6	3.98±0.35	4.2±0.3 [44] 4.2 [1,6]	4.4±0.3 [44] 4.7 [6] 4.7 [152]	4.29±0.74	4.58±0.50 [31]
I¹³¹	8.08 days	—	—	4.3 [1,6] 0.83±0.05 [113]	0.91±0.05 [113] 4.8 [6] 2.7±0.2 [152]	3.55±0.59	—
I¹³²	2.26 h	—	—	5.0 [1,6] 0.8 [88]	4.5±0.4 [152]	—	—
I¹³³	20.8 h	—	—	5.4 [1,6]	2.6±0.3 [152]	—	—
I¹³⁴	52.5 min	—	—	5.3 [1,6] 2.5 [88]	4.7±0.5 [152]	—	—
I¹³⁵	6.68 h	—	—	4.5 [1,6]	5.0±0.5 [152]	—	—
Xe¹³¹	Stable	—	—	(4.3)‡ [1,6]	—	—	—
Xe¹³²	Stable	—	—	(5.0)‡ [1,6]	—	—	—
Xe¹³³	5.27 days	—	—	—	6.6 [6]	—	—
Xe¹³⁴	Stable	—	—	(5.9)‡ [1,6]	—	—	—
Xe¹³⁵	9.13 h	—	—	—	5.5 [6]	—	—
Cs¹³³	Stable	—	—	(5.6)‡ [1,6]	—	—	—
Cs¹³⁵	3.0·10⁶ yr	—	0.5	(5.7)‡ [1,6]	—	—	—
Cs¹³⁶	12.9 days	—		0.145±0.020 [44] ~0.2 [88] 0.23 [6]	0.034±0.004 [44]	—	—
Cs¹³⁷	26.6 yr	—	4.7±0.5	5.9±0.6 [44]	6.6±0.6 [44]	—	5.1±0.8 [31]
Ba¹³⁹	84.0 min	—	—	5.0 [1,6]	4.6 [6] 4.4±0.5 [152]	4.84±0.35	—

TABLE 14 (Continued)

Nuclide	Half-life	Th²³² [21]	U²³³ [31]	U²³⁵ [1, 6, 44, 88*, 113*—115]	U²³⁸ [6, 44, 94, 108, 113†, 115, 116, 152]	Np²³⁷ [117]	Pu²³⁹ [31, 113†]
Ba¹⁴⁰	12.8 days	—	—	4.2±0.3 [44] 4.7 [1] 0.86±0.04 [113] 4.6 [6]	4.9±0.4 [44] 4.3±0.4 } [152] 4.6 4.41±0.22 [116] 0.80±0.04 [113] 4.6 [6]	4.89±0.35	0.64±0.03 [113] 4.35±0.40 [31]
Ce¹⁴¹	33.1 days	5.9±0.8	5.0±0.5	3.8±0.4 [44]	5.8±0.6 [44]	—	—
Ce¹⁴³	33 h	—	—	3.9 [1,6]	3.91±0.27 [116] 3.6 [6]	3.60±0.74	—
Ce¹⁴⁴	285 days	—	—	3.3 [1,6]	2.68±0.16 [116] 3.3 [6] 3.4 [152]	—	—
Pr¹⁴³	13.76 days	—	—	—	3.16±0.16 [116] 3.2 [6]	—	—
Nd¹⁴⁷	11.06 days	—	—	—	1.99±0.10 [116] 2.0 [6]	1.73±0.25	—
Sm¹⁵³	47.1 h	—	—	—	0.39±0.02 [116] 0.39 [6]	0.32±0.025	—
Eu¹⁵⁶	15.4 days	—	—	0.055 [1,6]	0.13±0.01 [116] 0.12 [6] 0.22 [152]	—	—
Eu¹⁵⁷	15.4 h	—	—	—	—	0.094±0.030	—
Gd¹⁵⁹	18.0 h	—	—	—	—	0.069±0.030	—

TABLE 15. Mo^{99} Yields in Fission of U^{235} and U^{238} by Neutrons (%) [94]

Fissionable nuclides	Neutron energy, MeV				
	Thermal	0.95	1.55	4.85	14.2
U^{235}	6.14±0.16	6.10±0.16	—	5.45±0.16	—
U^{238}	—	—	6.19±0.15	6.45±0.16	5.68±0.14

TABLE 16. Ratio of Yields of Mo^{99}, Zr^{97}, Ag^{113}, Cd^{155} in Fission of U^{235} by Neutrons of Various Energies [149]

Neutron energy, keV	$\dfrac{Y_{Mo^{99}}}{Y_{Ag^{113}}}$	$\dfrac{Y_{Zr^{97}}}{Y_{Mo^{99}}}$	$\dfrac{Y_{Zr^{97}}}{Y_{Cd^{115}}}$
Thermal	551±11	0.974±0.005	625±8
65^{+30}_{-25}	956±186	—	—
125^{+5}_{-35}	803±107	1.08±0.05	809±61
200^{+5}_{-38}	458±87	...	—
300^{+5}_{-42}	458±59	—	—
540^{+20}_{-68}	392±44	—	—
1000^{+20}_{-87}	404±46	—	—
14000	4.41±0.48	—	—

TABLE 17. Ratio of Relative Yields of Fission Fragments in the Fission of U^{235} and U^{238} by Neutrons of from 2 to 10 MeV to the Relative Yields of the Same Fragments in the Fission of U^{235} by Thermal Neutrons $\left(\dfrac{Y_i / Y_{Mo^{99}}}{Y_i^{Therm} / Y_{Mo^{99}}^{Therm}} \right)$ [151]

Fissionable nuclide	Neutron source	Sr89	Y91	Pd109	Ag111	Pd112	Cd115	Cs136	Cs137	Ba140	Nd147	Sm153	Eu156	Tb161
U235	p (12MeV) + Be9	0.85	1.08	4.07	4.16	5.12	4.72	3.22	0.97	0.94	0.89	1.25	1.80	—
		0.83	1.03	4.09	4.34	5.47	4.69	2.91	0.87	0.94	0.90	1.20	1.72	6.59 ± 0.46
		—	1.03 ± 0.07	4.15	3.87	5.05	4.62	3.12	1.03	0.94	0.98	1.27	1.76	6.96
	d (24 MeV) + Cu63,65	0.81	—	—	13.4 ± 2.0	—	18.1	9.99	0.98	0.87	0.88	1.37	2.43	18.2
				14.9	17.6	23.4	22.9			1.02				
	α (48MeV) + Al27	0.88	1.02	14.1	19.6	26.2	23.6	14.2	1.03	0.91	0.93	1.47	2.75	23.7
		0.77 ± 0.07	1.00	14.0	17.4	24.6	22.2	13.3	1.03	0.92	0.92	1.42	2.69	24.0
	α (48MeV) + Be9	0.83	1.02	—	26.1 ± 1.6	—	32.2	18.6	1.06	0.93	—	1.46	2.99	30.8
	d (24 MeV) + Be9	0.84	1.06	21.4	26.5	34.1	33.6	18.5	1.01	0.86	0.88	1.48	3.00	32.1
	d (24MeV) + Li6,7	0.87	1.03	23.8	31.5	42.4	41.7	21.3	0.98	0.92	0.92	1.60	3.43	39.4
		0.91	1.10	23.1 ± 2.3	33.1	42.3 ± 4.2	38.9	22.9	1.10	0.99	0.96	1.62	3.47	39.2 ± 2.7
		—	1.15 ± 0.23	29.0 ± 5.8	38.1 ± 7.6	49.1 ± 9.4	46.9 ± 9.4	27.0 ± 5.4	1.26 ± 0.25	1.07 ± 0.21	1.09 ± 0.22	1.79 ± 0.36	3.76 ± 0.75	45.7 ± 9.1

p(12 MeV)+Be⁹	0.50	0.73	9.28	6.88	7.66	5.23	0.16±0.03	0.82	0.89	1.06	2.54	5.46	22.6±2.5
	—	0.75	9.63	5.88	7.28	5.05	0.12±0.03	0.94	0.88	1.11	2.57	5.47	22.7
d(24 MeV)+Cu⁶³·⁶⁵	0.52	0.76	—	18.0±1.8	—	21.4	1.40±0.10	0.94	0.84	1.02	2.64	6.37	48.3
	—	—	21.4	19.6	26.0	22.4	—	—	0.96	—	—	—	—
α(48 MeV)+Al²⁷	0.54	0.73	19.1	22.2	28.0	23.5	1.98±0.14	0.94	0.87	1.10	2.75	6.37	49.2±3.9
	0.54±0.05	0.81	23.4	20.0	29.7	24.1	2.06±0.14	1.02	0.95	1.14	2.86	6.95	58.1
α(48MeV)+Be⁹	0.54	0.74	—	30.0±1.8	—	32.5	3.07±0.21	0.92	0.87	0.95	2.70	6.52±0.39	64.2
d(24 MeV)+Be⁹	—	0.76	27.2	29.3	36.1	33.4	2.32±0.16	0.91	0.83	1.04	2.85	7.14	68.0
d(24 MeV)+Li⁶·⁷	0.56	0.72±0.06	28.5	33.8	42.6	37.7	3.15±0.22	0.91	0.87	1.08	2.67±0.24	7.06±0.64	78.2±7.0
	0.57	0.76	28.6±2.9	33.1	39.7±4.0	35.7	3.10±0.22	0.93	0.91	1.09	2.97	7.35	72.6
	—	0.75	30.5	35.6	42.6	39.0	3.57±0.25	0.99	0.93	1.14	2.97	7.59	80.0

U²³⁸

Remark: Errors are 5% unless the contrary is indicated.

TABLE 18. Independent and Cumulative Fragment Yields in the Fission of Neutrons (ratio to

Nu-clide	Half-life	U²³³ [4, 88. 97, 104, 147]		U²³⁵ [4, 88, 91,
		Independent	Cumulative	Independent
As⁷⁸	91 min	—	—	$(8.5\pm2.5)\cdot10^{-3}$ [147] 0.09 [118] 0.09 [88]
Br⁸²	35.87 h	$(1.1\pm0.5)\cdot10^{-3}$ [147]	—	$\sim 6\cdot10^{-4}$ ⎫ 1.4·10⁻⁴ ⎬[118] 1.6·10⁻⁴ ⎭ $1.5\cdot10^{-4}$ [88] $(2\pm1)\cdot10^{-4}$ [147]
Rb⁸⁶	18.66 days	$7\cdot10^{-5}$ [147]	—	1.2·10⁻⁵ ⎫ [118] 1.5·10⁻⁶ ⎭ 1.3·10⁻⁵ [88] 1.5·10⁻⁵ [147]
Kr⁸⁹	3.18 min	—	—	—
Kr⁹⁰	33 sec	—	—	—
Kr⁹¹	9.8 sec	—	—	—
Kr⁹²	3.0 sec	—	—	—
Kr⁹³	2.0 sec	—	—	—
Kr⁹⁴	1.4 sec	—	—	—
Kr⁹⁵	Short-lived	—	—	—
Kr⁹⁷	~1 sec	—	—	—
Y⁹⁰	64.2 h	$<8.0\cdot10^{-5}$ [88, 147]	—	$<3.2\cdot10^{-4}$ [91] <3·10⁻⁴ ⎫ <5·10⁻⁴ ⎭[118] $<8\cdot10^{-5}$ [88. 147]
Y⁹¹	57.5 days	—	—	$<9.3\cdot10^{-3}$ [91] $<9\cdot10^{-3}$ [118]
Nb⁹⁶	23.35 h	$(1.0\pm0.2)\cdot10^{-3}$ [97] $(1.3\pm0.2)\cdot10^{-3}$ [147]	—	9·10⁻⁵ ⎫ [118] 1.4·10⁻⁴ ⎭ 9.0·10⁻⁵ [88] $(1.0\pm0.2)\cdot10^{-4}$ [147] $(9.8\pm2.0)10^{-5}$ [97]

97, 104, 118, 147]	Pu^{239} [4, 88, 97, 104, 147]		Cm^{242} [88]	Th^{232} [4]
Cumulative	Independent	Cumulative	Independent	Independent
—	—	—	—	—
—	—	—	—	—
—	$3.1 \cdot 10^{-5}$ [147] $1.6 \cdot 10^{-4}$ [88]	—	—	—
0.960 ± 0.004 [118, 147]	—	—	—	—
0.86 ± 0.02 [118, 147]	—	—	—	—
0.59 ± 0.01 [118]	—	—	—	—
0.31 ± 0.01 [118, 147]	—	—	—	—
$0.075 \, {}^{+\,0.010}_{-\,0.002}$ [147]	—	—	—	—
$0.015 \, {}^{+\,0.005}_{-\,0.002}$ [147]	---	—	—	—
$(11 \, {}^{+3}_{-1}) \cdot 10^{-4}$ [147]	--	—	—	—
$<1 \cdot 10^{-5}$ [147]	—	—	—	—
—	—	—	--	—
—	—	—	—	—
—	$(7.7 \pm 1.0) \cdot 10^{-4}$ [147] $(6.2 \pm 1.0) \cdot 10^{-4}$ [97]	—	—	—

TABLE 18

Nu-clide	Half-life	U²³³ [4, 88, 97, 104, 147]		U²³⁵ [4, 88, 91,
		Independent	Cumulative	Independent
Nb⁹⁷	72.1 min	$(1.1\pm0.4)\cdot10^{-2}$ [147]	—	$(1.7\pm0.8)\cdot10^{-3}$ [147]
Te⁹⁸	$1.5\cdot10^6$ yr	—	—	$3\cdot10^{-8}$ [118] $(3\pm2)\cdot10^{-8}$ [147]
Rh¹⁰²	210 days	—	—	$<2\cdot10^{-7}$ [118, 147]
Te¹³¹	24.8 min	—	—	0.14 $\left.\begin{array}{l}0.04\div0.12\\0.15\pm0.07\end{array}\right\}$ [118] 0.12 [147]
Te¹³¹ᵐ	30 h	0.23 [147]	—	—
Te¹³²	77.7 h	—	—	0.36 ± 0.17 [118, 147]
I¹²⁸	24.99 min	$(1.1\pm0.1)\cdot10^{-3}$ [88] $(1.0\pm0.1)\cdot10^{-4}$ [147]	—	$1.0\cdot10^{-4}$ [118] $(9.8\pm0.2)\cdot10^{-5}$ [88, 147]
I¹³⁰	12.6 h	$3.8\cdot10^{-3}$ [88] $(1.3\pm0.2)\cdot10^{-3}$ [147]	—	$(2.8\pm0,2)\cdot10^{-4}$ [88, 147] $2.8\cdot10^{-4}$ [118]
I¹³¹	8.08 days	—	—	<0.01 [118, 147]
I¹³²	2.26 h	—	—	<0.01 [88, 118, 147]
I¹³³	20,8 h	—	—	<0.05 [118, 147]
I¹³⁴	52.5 min	—	—	$\left.\begin{array}{l}0.13\\0.11\end{array}\right\}$ [118] 0.21 [88] $0.12\pm0,02$ [147]
I¹³⁶	86 sec	—	0.27 ± 0.005 [147]	—
Xe¹³³	5.270 days	—	—	<0.001 [118, 147]
Xe¹³⁵	9.13 h	—	—	$\left.\begin{array}{l}0.035\\0.049\\0.027\end{array}\right\}$ [118] 0.036 [88] 0.037 ± 0.012 [147]
Xe¹³⁶	Stable	—	—	0.57 [88]
Xe¹³⁷	3.9 min	—	—	(0.40) [88]

97, 104, 118, 147]	Pu²³⁹ [4, 88, 97, 104, 147]		Cm²⁴² [88]	Th²³² [4]
Cumulative	Independent	Cumulative	Independent	Independent
—	$(1.5\pm0.4) \cdot 10^{-2}$ [147]	—	—	—
—	—	—	—	—
—	—	—	—	—
—	—	—	—	—
—	0.18 [147]	—	—	—
—	—	—	—	—
—	$(2.4\pm0.2) \cdot 10^{-4}$ [88, 147]	—	—	—
—	$(2.0\pm0.1) \cdot 10^{-3}$ [88, 147]	—	—	—
—	—	—	—	—
—	—	—	—	—
—	—	—	—	—
—	—	—	—	—
0.48 ± 0.08 [147]	—	0.32 ± 0.06 [147]	—	—
—	—	—	—	—
—	—	—	—	—
—	—	—	—	—
0.978 ± 0.003 [147]	—	—	—	—

TABLE 18

Nu-clide	Half-life	U²³³ [4, 88, 97, 104, 147]		U²³⁵ [4, 88, 97,
		Independent	Cumulative	Independent
Xe¹³⁸	17 min	—	—	—
Xe¹³⁹	41 sec	-·-	—	—
Xe¹⁴⁰	16.0 sec	—	—	—
Xe¹⁴¹	1.7 sec	—	—	—
Xe¹⁴³	1.0 sec	—	—	—
Xe¹⁴⁴	∼1 sec	—	—	—
Cs¹³⁶	12.9 days	2.0·10⁻² [88] 1.70·10⁻² [104] (1.4±0.4)·10⁻² [147]	—	1.0·10⁻³ } [118] 9·10⁻³ 1.0·10⁻³ [88] 9.4±0.4·10⁻⁴ [147] 1.06·10⁻³ [104]
La¹⁴⁰	40.22 h	(3.8±0.1)·10⁻³ [88, 147]	-·-	<0.03 [118] ' 7.0·10⁻⁴ [88]
La¹⁴¹	3.8 h	—	—	∼0.02 [118] 2.0·10⁻² [88]
Pm¹⁴⁸	5.3 days	—	—	<10⁻⁴ [118, 147]
Pm¹⁵⁰	2.7 h	—	—	(2.1±0.1)·10⁻³ [147]

TABLE 19. Relative Yields of Krypton Isotopes in the Fission of Th²³², U²³⁵, 25.5 MeV

Relative yields	Th²³²			U²³⁵	
	Deutrons (MeV)		α-Particles	Deutrons (MeV)	
	9.3	14.0	25.5 MeV	9.3	14.0
$\dfrac{Y_{Kr^{85m}}}{Y_{Kr^{88}}}$	0.649± ±0.024	0.643± ±0.019	0.553± ±0.004	0.488± ±0.031	0.534± ±0.008
$\dfrac{Y_{Kr^{87}}}{Y_{Kr^{88}}}$	0.941± ±0.034	0.903± ±0.024	0.869± ±0.011	0.710± ±0.118	0.797± ±0.007

Continued)

104, 118, 147]	Pu²³⁹ [4, 88, 97, 104, 147]		Cm²⁴² [88]	Th²³² [4]
Cumulative	Independent	Cumu-lative	Inde-pendent	Inde-pendent
0.956±0.003 [147]	—	—	—	—
0.82±0.02 [118]	..	—	..	—
0.59±0.01 [118]	—	---	—	—
0.21±0.02 [118]	—	—	—	—
(8.5±0.5) · 10⁻³ [118]	—	—	—	—
(1.1±0.1) · 10⁻³ [118, 147]	—	—	—	—
—	3.1 · 10⁻⁸ [88] 1.66 · 10⁻² [104] (1.0±0.7) · 10⁻² [147]	—	0.11	<2.8 · 10⁻⁴
—	—	—	—	—
—	—	—	—	—
—	—	—	—	—
—	—	—	—	—

U²³⁸ by Deuterons of 9.3 and 14.0 MeV Energies, and α-Particles of
Energy [154]

α-Particles	Ther-mal	U²³⁸		
		Deutrons (MeV)		α-Particles
25.5 MeV	neutrons	9.3	14.0	25.5 MeV
0.596± ±0.046	0.386± ±0.025	0.547± ±0.005	0.552± ±0.006	0.576± ±0.016
0.920± ±0.018	0.659± ±0.013	0.921± ±0.009	0.878± +0.013	0.942± ±0.033

TABLE 20. Independent Fragment Yields in the Fission of U^{235} and U^{238} by 14 MeV Neutrons (ratios to total chain yields)

Nuclide	Half-life	U^{235} [88, 119]	U^{238} [120]
Nb^{96}	23.35 h	<0.01 [119]	—
Nb^{97}	72.1 min	0.069±0.015 [119]	—
Ag^{112}	3.20 h	0.051±0.006 [119]	—
Te^{131m}	30 h	0.35 [88]	—
I^{132}	2.26 h	0.16 [88]	—
I^{134}	52.5 min	0.43 [88]	—
Cs^{136}	12.9 days	0.04 [88] 0,04±0,005 [119]	—
Cs^{139}	9.5 min	—	0.286±0,015
Ba^{139}	84.0 min	—	0,056±0,013

3. FISSION BY GAMMA RAYS

TABLE 21. Fragment Yields (%) in the Fission of Bi^{209}, Ra^{226}, Th^{232}, and U^{235} by Gammas of Various Energies

Nuclide	Half-life	Bi^{209}, γ-Rays 85 MeV [121]	Ra^{226}, γ-Rays 23 MeV [5]	Th^{232}, γ-Rays 10 MeV [123]	Th^{232}, γ-Rays 69 MeV [122]	U^{235}, γ-Rays 20 MeV [5]
Ga^{73}	5 h	—	<0.4	—	—	—
Ge^{77}	11.3 h	~0.3	—	—	—	—
Ag^{77}	38.7 h	<0.4	—	—	—	—
Br^{82}	36 h	<0.5	—	—	—	—
Br^{83}	2.3 h	1.2	4.1±0.9	1.8±0.44	1.89±0.15	—
Br^{84}	31.8 min	<1.4	—	—	—	—
Sr^{89}	50.5 days	—	—	—	6.7±0.1	4.9
Sr^{91}	9.67 h	2.8	3.4±0.9	5.7±0.91	5.7±0.1	5.4
Sr^{92}	2.6 h	2.8	3.4±0.9	—	—	—
Zr^{97}	17 h	3.0	1.4±0.7	2.3±0.51	—	6.1*
Mo^{99}	66 h	—	—	1.1±0.2	1.85±0.10	5.2
Ru^{105}	4.5 h	5.0†	2.4±0.6	—	0.83±0.07	—
Pd^{109}	13.5 h	~6.4	—	—	—	—
Ag^{111}	7.6 days	~2.8	—	—	0.90±0.09	—
Ag^{112}	3.2 h	—	—	—	0.68±0.02	—
Ag^{113}	5.3 h	3.0	4.2±2.2	0.066±0.016	0.58±0.01	—
Cd^{115}	53 h	—	—	0.032±0.008	—	0.27
Cd^{117}	~50 min	—	3.1±1.2	0.037±0.011	0.68±0.02	—
Sb^{127}	88 h	—	—	—	—	2.3
Sb^{129}	4.2 h	—	<2.1	0.50±0.25	—	—
I^{131}	8.08 days	—	—	—	2.25±0.10	—
I^{133}	20.8 h	—	—	4.3±1.7	—	—
I^{134}	52.5 min	<0.2	—	—	—	—
I^{135}	6.7 h	—	4.0‡	—	—	—
Ba^{139}	84 min	<0.1	—	5.0 0.75	—	—
Ba^{140}	12.8 days	—	—	7. 71.5	6.6±0.5	—
Ce^{141}	33.1 days	—	—	—	6.8±0.5	—
Ce^{143}	33 h	—	2.1±2.0	9.5	4.8 0.5	—
Ce^{144}	285 days	—	—	—	4.8	4.7

* All yields normalized to the yield of Zr^{97}, 6.1%.
† All yields normalized to the yield of Ru^{105}, 5%.
‡ All yields normalized to the yield of I^{135}, 4%.

TABLE 22. Fragment Yields in the Photofission of Natural Uranium (%)

γ-Ray energy, MeV

Nuclide	Half-life	5.5 [5]	6.0 [5]	6.5 [5]	7.0 [5,124]	7.5 [5]	8.0 [5]	10 [5,124,125]	12 [5,126]	16 [5,124,125]	18 [5,126]	21 [5,124]	22 [5,126]	31 [5]	48 [124,125]	100 [5,124]	300 [5]
Ge^{77}	11.3 h	—	—	—	—	—	—	—	—	—	—	—	—	—	0.036	—	—
As^{77}	38.7 h	—	—	—	—	—	—	—	—	—	—	—	0.05	—	—	—	—
Ge^{78}	86 min	—	—	—	—	—	—	—	—	—	—	—	—	—	0.056	—	—
Br^{83}	2.3 h	0.25	0.27	0.26	0.25	—	0.32	0.30	0.37	0.29	0.45	—	0.47	0.47	0.70	0.74	0.87
Br^{84}	31.8 min	0.63	0.58	0.60	0.60	—	0.72	0.41	—	0.51	—	—	—	—	1.65	1.66	1.75
Sr^{89}	50.5 days	—	—	2.8	3.3	3.2	3.1	4.4	—	3.7	—	2.8	4.0	3.3	3.0	3.0	3.2
Sr^{91}	9.7 h	3.4	—	3.8	—	—	3.9	—	—	4.2	4.2	—	—	3.9	3.9	—	—
Sr^{92}	2.6 h	—	—	—	—	—	—	—	—	3.5	—	—	—	—	—	—	—
Y^{93}	10.4 h	—	—	—	—	—	—	5.1	—	5.3	6.1	—	5.6	5.3	—	—	—
Zr^{97}	17 h	—	—	—	6.4	—	—	4.9	—	6.3, 6.1	5.2	5.5	5.3	5.0	5.3	5.6	—
Mo^{99}	66 h	6.1	—	—	6.4	—	6.4	5.6	—	5.6	—	5.6	—	—	5.6	5.6	5.6
Ru^{103}	39.8 days	—	—	—	—	—	3.0	—	—	3.6	—	3.7	3.0	—	3.6	4.0	4.2
Ru^{105}	4.5 h	3.9	4.3	3.6	—	—	2.0	—	2.5	—	2.0	—	2.0	—	2.2	—	—
Ru^{106}	1 yr	—	—	—	—	—	—	—	—	—	—	1.5	—	—	1.4	1.8	2.1
Rh^{105}	36.5 h	—	—	—	—	—	—	—	—	—	—	—	—	2.1	—	—	—
Pd^{109}	13.5 h	—	—	—	—	—	0.11	0.085	—	0.22	—	—	—	—	—	—	—

Isotope	Half-life															
Pd^{112}	21 h	—	—	—	0.033	0.017	0.042 0.045	—	0.11 0.15	—	0.25	—	0.37	0.49	0.68	1.08
Ag^{111}	7.6 days	—	—	0.032	0.025 0.050	0.025	0.062	—	0.28	0.20	0.41	0.27	0.46	0.73	0.97	1.79
Ag^{113}	5.3 h	—	—	—	—	—	—	—	0.063	—	—	—	0.31	0.57	0.73	1.15
Cd^{115m}	43 days	—	—	0.024	0.021	0.018	—	} 0.090	0.012 0.052	} 0.25	0.017	} 0.330	0.025	0.039	0.046	0.19
Cd^{115}	53 h	—	—	—	—	0.022	0.029	}	0.15	}	0.24	}	0.31	0.44	0.64	0.09
Cd^{117}	3 h	0.030	0.054	0.025	0.022	0.022	0.026	—	—	—	—	—	0.32	0.48	0.66	0.99
Sn^{121}	27.5 h	—	—	—	—	—	—	—	—	—	—	—	0.30	—	—	—
Sn^{125}	9.4 days	—	—	—	—	—	—	—	—	—	—	—	0.40	—	—	—
Sb^{127}	88 h	—	—	—	0.062	0.052	—	—	—	—	1.8	—	0.76	2.4	2.7	3.8
Sb^{129}	4.2 h	0.27	—	—	—	0.30	—	—	—	—	—	—	1.4	—	—	—
Te^{132}	77.7 h	—	—	—	—	—	5.6	—	5.8	—	4.9	—	3.9	4.8	4.5	4.2
I^{131}	8.08 days	—	—	—	—	—	3.8	—	4.4	—	4.0	—	2.6	4.3	4.3	4.6
I^{133}	20.8 h	—	6.0	—	—	—	6.8	—	7.1	—	—	—	5.8	6.0	—	—
Cs^{137}	26.6 yr	—	—	—	—	—	—	—	—	—	—	5.3	—	4.3	—	—
Cs^{138}	32.2 min	6.0	6.0	6.0	6.0	6.0	5.9	6.0	6.0	5.1	—	4.8	—	—	—	—
Ba^{139}	84 min	—	—	6.0	6.0 6.5	6.0	5.8 5.6	—	5.8	4.8	—	—	—	4.3	4.8	—
Ba^{140}	12.8 days	—	—	—	—	—	6.0	—	4.9	—	4.8	4.4	4.8	4.8	5.2	4.7
Ce^{141}	33.1 days	—	—	—	—	—	—	—	—	4.5	—	—	4.5	4.5	—	—
Ce^{143}	33 h	4.1	—	—	—	4.1	—	—	5.3	—	4.4	4.4	4.3	4.2	4.2	4.0
Ce^{144}	285 days	—	—	—	—	4.3	—	—	—	—	4.7	—	—	4.2	—	—

4. FISSION BY CHARGED PARTICLES

TABLE 23. Fragment Yields in Fission of Ra^{226}, Bi^{209}, and U^{238} by Protons of Various Energies

Nuclide	Half-life	Ra^{226} (Yield in %) 11 MeV [41]	Bi^{209} (Yield in μbarn)			U^{238} (Yield in mbarn)			
			36 MeV [127]	58 MeV [127]	75 MeV [128]	10 MeV [129]	32 MeV [129]	70 MeV [129]	100 MeV [129]
Ni^{66}	56 h	—	0.48±0.08	—	—	—	—	—	—
Cu^{67}	61 h	—	0.72±0.14	5.5±0.8	—	—	—	—	—
Zn^{71m}	3 h	—	0.80±0.04	9.5±3.8	—	—	—	—	—
Zn^{72}	49 h	—	0.85±0.05	15.4±2.9	—	—	—	—	—
Ga^{72}	14.8 h	—	0.15±0.01	2.5	—	—	—	—	—
Ga^{73}	5 h	—	0.44±0.03	17	—	—	—	—	—
As^{74}	17.5 days	—	—	—	9	—	—	—	—
As^{77}	38.7 h	—	0.76±0.13	35±5	84	—	—	—	—
Br^{82}	36 h	—	0.26±0.03	5.9±0.9	38	—	—	—	—
Br^{83}	2.3 h	3.1±0.2	5.7±1.0	123±8	39	—	—	—	—
Br^{84}	31.8 min	2.9±0.2	—	—	26	—	—	—	—
Rb^{86}	18.66 days	—	2.7	5.1	—	—	—	—	—
Sr^{89}	50.5 days	—	35±5	318±12	—	—	—	—	—
Sr^{91}	9.67 h	3.5±0.4	56±3	490±20	—	—	—	—	—
Sr^{92}	2.6 h	3.0±0.4	—	—	—	—	—	—	—
Y^{90}	64.2 h	—	—	—	—	—	0	0.02	0.11

Isotope	Half-life								
Y^{91}	57.5 days	—	—	—	—	0.34	27	27	30
Y^{93}	10.4 h	3.4	—	—	—	0.80	43	—	49
Zr^{97}	17 h	1.9±0.3	—	—	14	—	—	—	—
Nb^{95m}	90 h	—	—	—	19	—	—	—	—
Nb^{95}	35 days	—	—	—	30	—	—	—	—
Nb^{96}	23.35 h	—	—	—	—	1.9	62	71	69
Mo^{99}	66 h	1.9±0.2	167±3	790±50	—	—	—	—	—
Ru^{105}	4.5 h	2.9±0.4	—	—	—	—	—	—	—
Pd^{109}	13.5 h	3.6±0.4	—	—	—	—	—	—	—
Pd^{112}	21 h	3.8±0.3	—	—	—	—	—	—	—
Ag^{111}	7.6 days	4.0	93±9	590±60	400	—	—	—	—
Ag^{112}	3.2 h	—	88±9	520±80	130	—	—	—	—
Ag^{113}	5.3 h	3.1±0.1	85±13	380±40	320	—	—	—	—
Cd^{115}	53 h	3.9±0.6	—	—	—	—	—	—	—
In^{111}	2.8 days	—	—	—	140	—	—	—	—
In^{114m}	50 days	—	—	—	5	—	—	—	—
In^{115m}	4.5 h	—	—	—	140	—	—	—	—
Sn^{121}	27.5 h	2.0±0.2	—	—	—	—	—	—	—
Sn^{125}	9.5 min	0.9±0.1	—	—	—	—	—	—	—
Te^{132}	77.7 h	2.8±0.5	—	—	—	—	—	—	—
I^{124}	4.5 days	—	—	—	0.34	—	—	—	—
I^{126}	2.6 h	—	—	—	1.4	—	—	—	—
I^{128}	24.99 min	—	—	—	1.4	—	—	—	—

TABLE 23   (Continued)

Nuclide	Half-life	Ra226 (Yield in %) 11 MeV [41]	Bi209 (Yield in μbarn) 36 MeV [127]	58 MeV [127]	75 MeV [128]	U238 (Yield in mbarn) 10 MeV [129]	32 MeV [129]	70 MeV [129]	100 MeV [129]
I130	12.6 h	—	—	—	0.94	—	—	—	—
I132	2.26 h	—	—	—	0.68	—	—	—	—
I134	52.5 min	4.7±0.4	—	—	—	—	—	—	—
Ba139	84 min	4.0	—	—	—	—	—	—	—
Ba140	12.8 days	2.8±0.2	—	—	—	—	2.8	8.6	8.6
La140	40.22 h	—	—	—	—	1.7	48	49	51
Ce141	33.1 days	—	—	—	—	1.2	45	36	31
Ce143	33 h	1.9±0.2	—	—	—	0	0	0.36	2.1
Pr143	13.76 days	—	—	—	—	1.1	35	30	28
Ce144	285 days	—	—	—	—	—	—	0	0
Nd140	3.3 days	—	—	—	—	0.60	—	0	0
Nd147	11.06 days	—	—	—	—	0.12	17	17	18
Sm153	47.1 h	—	—	—	—	0.02	4.2	4.6	4,4
Eu156	15.4 days	—	—	—	—	0.02	1.12	1.22	1.31
Eu157	15.4 h	—	—	—	—	0.02	0.69	0.90	0.89

TABLE 24. Relative Fragment Yields in the Fission of Ra[226] by Protons of Various Energies (with respect to yield in fission by protons of 13.6 MeV energy) [130]

Nuclide	Half-life	Proton energy, MeV			
		11.0	13.6	20.3	22.1
Sr91	9.67 h	0.15±0.01	1.00	3.90±0.117	4.26±0.35
Pd112	21 h	0.13±0.01	1.00	6.33±0.38	7.53±0.79

TABLE 25. Ratio of Yields of Ga[72], Br[82], Ag[112], and La[140] to Total Yields of the Corresponding Chains in the Fission of Th[232] by Protons of Various Energies (%) [153]

Nuclide	Half-life, h	Proton energy, MeV								
		12.8	17.2	18.6	25.8	36.1	39.5	40.7	55.5	82.4
Ga72	14.3	—	—	—	—	2.1	—	3.3	6.8	13.2
Br82	35.87	0.43	—	0.67	—	—	3.5	—	6.2	10.9
										12.5
Ag112	3.20	...	1.8	—	3.5	—	—	5.8	7.7	15.4
									8.9	15.8
La140	40.22	—	—	—	4.3	—	—	—	22	22

TABLE 26. Fragment Yields in the Fission of Natural Uranium by Deuterons

Nuclide	Half-life	5 (Yield in μbarn) [134]	Deuteron energy, MeV				
			10 (Yield in mbarn) [134]	10 (Yield with respect to Ba^{140}) [133]	13.6 (Yield in mbarn) [134]	13.6 (Yield with respect to Ba^{140}) [133]
Zn^{72}	49 h	—	—	—	0.026 ± 0.005	—
As^{77}	35.87 h	0.4	0.017	—	0.16 ± 0.04	—
Br^{82}	35.87 h	—	—	$9.4 \cdot 10^{-4}$	—	$(1.0 \pm 0.1) \cdot 10^{-3}$
Br^{83}	2.3 h	9	0.19 ± 0.04	0.15 ± 0.03	2.2 ± 0.1	0.19 ± 0.01
Br^{84}	31.8 min	—	—	0.21 ± 0.02	—	0.23 ± 0.01
Rb^{86}	18.66 days	—	—	$\ll 2.9 \cdot 10^{-4}$	—	$< (1.12 \pm 0.20) \cdot 10^{-4}$
Sr^{89}	50.5 days	94 ± 4	1.76 ± 0.06	0.52 ± 0.05	9.3 ± 0.4	0.61 ± 0.07
Y^{91}	57.5 days	143 ± 7	2.3 ± 0.1	—	12.2 ± 0.8	—
Y^{93}	10.4 h	180	2.9	—	16.4 ± 0.8	—
Zr^{95}	65 days	205 ± 15	3.5 ± 0.2	—	16.8 ± 0.7	—
Zr^{97}	17 h	233 ± 16	5.2 ± 0.2	—	24.1 ± 0.2	—
Ru^{103}	39.8 days	141 ± 2	2.6	—	13 ± 2	—
Ru^{105}	4.5 h	134 ± 3	2.6	—	15.7 ± 1.5	—
Ru^{106}	1.0 yr	120 ± 20	2.4	—	14.0 ± 0.7	—
Ag^{111}	7.6 days	—	—	0.30	—	0.37 ± 0.02
Ag^{112}	3.2 h	—	—	0.30	—	0.35 ± 0.03
Ag^{113}	5.3 h	—	—	0.32	—	0.25 ± 0.01

Isotope	Half-life					
Ag^{115}	21.1 min	—	—	0.30	—	0.27±0.02
Cd^{115}	53 h	4.8±0.5	0.73±0.03	—	4.2±0.2	—
Cd^{115m}	43 days	0.4±0.2	0.046±0.004	—	0.30±0.02	—
Sb^{127}	88 h	15±1	1.56±0.06	—	7.3	—
Sb^{129}	4.2 h	11	1.15±0.10	—	5.9	—
I^{131}	8.08 days	128±1	4.2±0.1	0.77	19±2	0.62±0.01
I^{132}	2.26 h	172±3	4.2±0.2	1.21	20.6±1.1	1.07±0.10
I^{133}	20.8 h	237±9	5.3±0.2	—	23	—
Cs^{136}	12.9 days	—	—	0.049±0.001	—	0.057±0.006
Cs^{137}	26.6 yr	—	—	1.52±0.03	—	1.33±0.16
Ba^{139}	84 min	—	—	1.22±0.10	—	1.24±0.05
Ba^{140}	12.8 days	181±9	3.4±0.2	1.05	17.5±0.9	1.05
Ce^{141}	33.1 days	160±16	2.9±0.1	—	15.6±1.6	—
Ce^{144}	285 days	183±4	3.0±0.3	—	12.1±0.3	—
Pr^{143}	13.76 days	120±7	2.6±0.1	—	14±2	—
Nd^{147}	11.06 days	89±4	1.6±0.1	—	8.7±0.3	—
Pm^{149}	54 h	33±2	0.8	—	3.8±0.6	—
Pm^{151}	27.5 h	—	0.5	—	3.0±0.2	—
Sm^{153}	47.1 h	12.1±0.1	0.33±0.01	—	2.0±0.1	—
Eu^{156}	15.4 days	3.0±0.5	0.132±0.002	—	0.66±0.08	—
Eu^{157}	15.4 h	1.6	0.076±0.002	—	0.45±0.06	—
Gd^{159}	18 h	—	0.039±0.001	—	0.21±0.01	—

TABLE 27. Fragment Yields in the Fission of Bi^{209}, Ra^{226}, and Th^{232} by Deuterons of Various Energies

Nuclide	Half-life	Bi^{209} [131] (Yield in %) 22 MeV	Ra^{226} [132] (Yield with respect to Ag^{111})		Th^{232} [133] (Yield with respect to Si^{89}) 13.6 MeV
			14.5 MeV	21.5 MeV	
Br^{82}	36 h	—	—	—	$(1.1\pm0.3)\cdot10^{-3}$
Br^{83}	2.3 h	—	2.4	1.4 ± 0.1	0.34 ± 0.02
Br^{84}	31.8 min	—	—	—	0.40
Rb^{86}	18.66 days	—	—	—	$<25\cdot10^{-4}$
Sr^{89}	50.5 days	—	—	—	1.00
Sr^{91}	9.67 h	2.0	3.1 ± 0.2	2.2 ± 0.2	—
Sr^{92}	2.6 h	2.1	2.3	2.1 ± 0.1	—
Y^{92}	3.6 h	—	3.0 ± 0.4	2.2 ± 0.5	—
Y^{93}	10.4 h	3.6	3.9 ± 0.4	2.8 ± 0.2	—
Zr^{97}	17 h	6.5 ± 1.4	2.2 ± 0.2	2.7 ± 0.2	—
Mo^{99}	66 h	10.2 ± 0.6	3.0 ± 0.2	3.8 ± 0.2	—
Ru^{105}	4.5 h	10.0 ± 1.2	4.5 ± 0.2	4.5 ± 0.2	—
Pd^{109}	13.5 h	6.4 ± 0.6	4.7 ± 0.2	4.4 ± 0.2	—
Pd^{111m}	5.5 h	1.6 ± 0.08	—	—	—
Pd^{112}	21 h	6.0 ± 0.5	5.3 ± 0.2	5.2 ± 0.1	—
Ag^{111}	7.6 days	5.1 ± 1.9	5.0	5 0	0.31
Ag^{112}	3.2 h	—	—	—	0.31
Ag^{113}	5.3 h	4.5 ± 0.08	3.9 ± 0.1	4.0 ± 0.1	0.25
Ag^{115}	21.1 min	—	—	—	0.27
Cd^{115}	53 h	—	4.6 ± 0.2	4.7 ± 0.1	—
Sb^{127}	88 h	<0.13	1.9	—	—
I^{131}	8.08 days	—	3.3 ± 0.2	2.6 ± 0.4	0.36
I^{132}	2.26 h	—	—	—	0.82
I^{133}	20.8 h	—	5.5 ± 0.2	3.3 ± 0.4	—
I^{134}	52.5 min	—	—	—	1.00
I^{135}	6.68 h	—	—	—	1.20
Cs^{136}	12.9 days	—	—	—	0.04 ± 0.01
Cs^{137}	26.6 yr	—	—	—	1.25 ± 0.02
Ba^{139}	84 min	$<0.06\pm0.02$	—	—	0.85
Ba^{140}	12.8 days	—	2.4	2.1 ± 0.3	0.87

TABLE 28. Fragment Yields in the Fission of U^{238} by Deuterons (in mbarn) [129]

Nuclide	Half-life	Energy, MeV				
		20	28	50	75	100
Y^{90}	64.2 h	—	0	<0.01	0.083	0.55
Y^{91}	57.5 days	19	20	22	38	42
Y^{93}	10.4 h	24	25	46	65	71
Mo^{99}	66 h	55	60	68	92	101
La^{140}	40.22 h	—	0.44	4.0	10	12
La^{141}	3.8 h	28	30	54	71	67
Ce^{141}	33.1 days	—	0	0	0	4.5
Ce^{143}	33 h	22	28	37	49	1.5
Ce^{144}	285 days	20	25	33	43	46
Pr^{143}	13.76 days	—	0	0	0.52	39
Nd^{140}	3.3 days	—	<0.2	<0.2	<0.2	<0.2
Nd^{147}	11.06 days	11	12	22	29	27
Sm^{153}	47.1 h	2.5	2.7	5.7	7.2	7.0
Eu^{156}	15.4 days	0.60	0.67	1.5	2.0	2.1
Eu^{157}	15.4 h	0.50	0.61	1.0	1.5	1.4

TABLE 29. Fragment Yields in the Fission of Ra^{226} and Th^{232} by Helium Nuclei

Nuclide	Half-life	Ra^{226} [134] (Yield relative to Ag^{111})			Th^{232} [135] (Yield in mbarn) 37.5 MeV
		23.5 MeV	31 MeV	43 MeV	
Zn^{72}	49 h	—	—	—	<0.08
As^{77}	38.7 h	—	—	—	~0.028
Br^{82}	35.87 h	—	—	—	<0.0185
Br^{83}	2.3 h	—	—	—	5.69 4.83
Sr^{89}	50.5 days	—	—	—	33.4 22.5
Sr^{90}	27.7 yr	—	—	—	21.6
Sr^{91}	9.67 h	10.8	6.6	5.2±0.4	20.8
Sr^{92}	2.6 h	8.6	5.6	4.4±0.2	22.3
Y^{92}	3.6 h	—	5.7	3.8+0.4	—
Y^{93}	10.4 h	—	5.0	5.2±0.2	—
Zr^{95}	65 days	—	—	—	19.4 17.9
Zr^{97}	17 h	—	3.7	4.2±0.2	—
Mo^{99}	66 h	5.0	4.3	5.4±0.2	18.1
Ru^{105}	4.5 h	—	4.1	5.2	—
Ru^{106}	1.0 yr	—	—	—	22.7
Pd^{109}	13.5 h	4.6	4.7	5.4	—
Pd^{112}	21 h	4 0	5.1	5.6	~16.5 8.3
Ag^{111}	7.6 days	5.0	5.0	5.0	13.2
Ag^{113}	5.3 h	3.9	3.9	4.2±0.2	—
Cd^{115}	53 h	4.4	4.5	5.0	14.3 13.2
Cd^{115m}	43 days	—	—	—	1.01 1.28

TABLE 29   (Continued)

Nuclide	Half-life	Ra^{226} [134] (Yield relative to Ag^{111})			Th^{232} [135] (Yield in mbarn) 37.5 MeV
		23.5 MeV	31 MeV	43 MeV	
Sn^{121}	27.5 h	—	—	—	6.2 10.9 9.8
Sb^{125}	2 yr	—	—	—	8.9 9.3
Te^{132}	77.7 h	—	—	—	14.4 9.1
I^{131}	8.08 days	—	—	4.0	9.8 6.8 8.2
I^{133}	20.8 h	—	—	5.6	—
Cs^{136}	12.9 days	—	—	—	0.3 0.31
Cs^{137}	26.6 yr	—	—	—	51.5
Ba^{140}	12.8 days	11.7	7.3	3.7 ± 0.5	24.4 24.7 19.6 17.4 17.2 22.3
Ce^{143}	33 h	—		—	16 6
Ce^{144}	285 days	—	—	—	16.7 17.3
Sm^{153}	47.1 h	—	—	—	< 6.3
Eu^{156}	15.4 days	—	—	—	0.25 0.35
Eu^{157}	15.4 h	—	—	—	0.31

TABLE 30. Fragment Yields in the Fission of U^{233} by Helium Nuclei (mbarn) [136]

Nuclide	Half-life	Energy, MeV								
		23.5	26.2	27.8	30.7	35.3	40.4	41.0	44.3	46.2
Sr^{89}	50.5 days	—	—	—	—	16	—	9 9	32	22
Sr^{91}	9.67 h	—	—	—	—	19	—	17	52	35
Zr^{95}	65 days	2.4	12	17	21	39	46	42	55	57
Zr^{97}	17 h	6.5	15	15	12	38	45	44	43	48
Mo^{99}	66 h	1.4	—	—	—	—	—	—	32	—
Ru^{103}	39.8 days	4.8	—	—	—	24	—	—	28	—
Ru^{105}	4.5 h	3.2	—	—	—	27	—	—	41	—
Ru^{106}	1.0 yr	—	—	—	—	27	—	—	45	—
Ag^{111}	7.6 days	>0.29	11	—	—	—	—	—	>44	—
Cd^{115}	53 h	3.3	9.9	15	32	41	—	43	68	74
Ba^{139}	84 min	4.6	9.2	—	—	18	—	12	22	25
Ba^{140}	12.8 days	3.4	7.4	—	—	12	—	8.4	14	16
Ce^{141}	33.1 days	10	—	26	13	—	39	—	—	—
Ce^{143}	33 h	8.4	—	13	12	—	—	—	—	—
Ce^{144}	285 days	2.1	—	8.0	8.8	—	15	3.0	—	—
Nd^{147}	11.06 days	2.0	—	8.0	—	—	15	—	—	—
Eu^{156}	15.4 days	—	—	0.68	—	—	0.44	—	—	—
Eu^{157}	15.4 h	0.04	—	0.94	—	—	0.48	—	—	—
Tb^{161}	6.88 days	—	—	0.50	—	—	0.71	—	—	—

TABLE 31. Fragment Yields in the Fission of U^{233} by
Helium Ions (mbarn) [137]

Nuclide	Half-life	Energy, MeV			
		25.3	29.0	34.5	40.5
Br83	2.3 h	3 4±0.1	5.1±0.1	9 5±0.5	8.6±0.1
Sr89	50.5 days	5.7±0.5	11.7±0.4	21.5±0.5	22.0±1.0
S r91	9.67 h	10.0±0.9	17.5±0.5	27.0±2.3	33 1±0.1
Sr92	2.6 h	11.4±0.4	23.5±0.9	28.5±0.5	34.8
Y93	10.4 h	17.0±0.4	23.3±0.3	35.9±1.0	38.8±0.4
Zr95	65 days	25.0±1.3	34.3±0.5	46.0±0.5	49.6
Zr97	17.0 h	21.0±1.9	31.0±1.9	45.9±1.0	46.0±2.5
Ru103	39.8 days	12.1±0.1	19.2±0.3	30.0±1.0	44.0±2.8
Ru105	4.5 h	7.6±0.1	11.0±0.9	22.3±0.3	35.0±1.2
Ru106	1.0 yr	10.4±0.6	15.0±0.5	31.0±0.9	40.0±3.2
Pd109	13.5 h	—	15.0	--	—
Pd112	21 h	7.0	13.0±0.3	—	—
Ag113	5.3 h	—	16.9	—	—
Cd115	53 h	8.0±0.5	16.0±0.9	33.2	42.1
Cd115m	43 days	0.8	(1.6)	4.0	5.3
I131	8.08 days	11.7±0.5	19.2	26.2	32.9
I133	20.8 h	9.0±0.5	12.0±1.3	14.7±0.2	13.5
Ba140	12.8 days	7.3±0.5	11.4±0.4	11.0	12.0
Ce141	33.1 days	12.2±1.0	15.5±1.1	19.3±0.7	16.3
Ce143	33 h	8.7±0.1	13.4±0.6	15.5	12.2
Pr145	5.95 h	6.0±0.1	10.4±0.1	13.0±0.2	10.3
Nd147	11.06 days	4.9±0.3	8.9±0.2	9.0	9.4±0.4
Sm153	47.1 h	1.0±0.10	1.9±0.05	2.7±0.3	2.8±0.1
Eu157	15.4 h	0.45±0.03	0.67	1.1±0.05	1.0
Gd159	18 h	0.20±0.02	0.27±0.05	0.55±0.05	0.55±0.05

TABLE 32. Fragment Yields in the Fission of U²³⁵ by Helium Nuclei (mbarn) [118]

Nuclide	Half-life	Energy, MeV						
		20.5	23.1	25.95	28.2	33.8	39.9
Zn⁷²	49 h	—	—	—	—	(0,085)	0.43
Br⁸³	2.3 h	—	0.66	—	3.12	5.3	7.6
Sr⁸⁹	50.5 days	0.38±0,02	2.12±0.03	5.9	11.3±0.7	17.6±1.4	20.6±0.1
Sr⁹¹	9.67 h	0.57±0.09	3.09±0.08	7.8	14.7±0.3	24.7±0.3	29.9±1.1
Sr⁹²	2.6 h	(0.47)	3.09±0.20	10.0	15.6±2.0	26.9±0.8	28.0±1.5
Y⁹³	10.4 h	—	—	—	—	27.6	(50)
Zr⁹⁵	65 days	0.56	5.0±0.15	—	25.0±1.0	39.0±0.8	48.0±2.6
Zr⁹⁷	17.0 h	0.83±0.03	5.0±0.19	15.3	26.5±1.7	41.0±2.3	49.0±1.2
Ru¹⁰³	39.8 days	—	3.15	—	22.5	37.7	48.5±0.6
Ru¹⁰⁵	4.5 h	(0.104)	2.0	—	17.8	25.0	39.8±1.8
Ru¹⁰⁶	1.0 yr	—	(2.0)	—	—	(31)	42.6±2.2
Pd¹¹²	21 h	—	—	5.9	11.4±2.6	28.8±2.2	40.8±2.3
Cd¹¹⁵	53 h	—	1.10	4.63	11.4	27.6±1.6	41.0±0.6

94

	Half-life						
Cd115m	43 days	—	—	—	—	—	5.0±0.15
I131	8.08 days	0.47	2.56	—	16.7±0.4	25.6±1.4	33.0±1.6
I133	20.8 h	0.72	3.5±0.3	—	18.4±0.7	23.7±0.7	23.9±1.6
Ba139	84 min	—	—	(9.07)	17.0	28.5	31.2
Ba140	12.8 days	0.57±0.10	3.4	8.70	13.5±0.5	20.5±0.7	22.0±0.2
La140	40.22 h	0.6¡	4.0	—	2.2±0.15	5.7±0.1	8.1±0.7
Ce141	33.1 days	0.61	—	—	17.5±0.7	24.0±1.0	32.0±0.9
Ce143	33 h	—	3.0±0.4	—	15.5±0.5	22.0±1.3	27.0±0.8
Pr145	5.95 h	—	—	—	—	17.5	23.4
Nd147	11.06 days	0,30	1.60±0.04	—	8.2±0.5	13.4±0.8	15.8±0.6
Sm153	47.1 h	0.05	0.44±0.02	—	2.7±0.4	4.0±0.06	5.2±0.14
Eu156	15.4 days	—	—	—	0,74	1.54±0.14	2.1±0.15
Eu157	15.4 h	0.0165	0.038	—	0.39	1.16±0.12	1.66±0.06
Gd159	18 h	—	—	—	0.29	0.68±0.03	1.05±0.06
Tb161	6.88 days	—	—	—	0.09	—	0.50

TABLE 33. Fragment Yields in the Fission of U^{235} by
Helium Nuclei (mbarn) [136]

Nuclide	Half-life	Energy, MeV								
		18.7	21.9	26.8	30.6	32.8	34.1	37.1	42.8	45
Zn^{72}	49 h	—	—	—	—	—	—	—	0.48	—
Sr^{89}	50.5 days	0.098	—	10	17	22	23	27	—	60
Sr^{91}	9.67 h	0.095	—	11	18	24	26	28	—	56
Zr^{95}	65 days	—	4.2	16	29	32	39	47	49	45
Zr^{97}	17.0 h	0.071	4.3	17	31	31	42	49	50	56
Mo^{99}	66 h	—	—	19	—	35	—	—	—	63
Ru^{103}	39.8 days	0.017	—	12	—	—	—	—	—	52
Ru^{105}	4.5 h	—	—	12	—	—	—	—	—	52
Ru^{106}	1.0 yr	—	—	11	—	—	—	—	—	60
Pd^{112}	21 h	—	—	10	—	30	—	—	—	48
Ag^{111}	7.6 days	—	—	11	—	33	—	—	—	72
Ag^{112}	3.2 h	—	—	—	—	—	—	—	—	48
Ag^{113}	5.3 h	—	—	11	—	—	—	—	—	76
Cd^{115}	53 h	—	0.87	12	18	36	38	48	51	60
Cd^{117}	~50 min		0.90	—	15	—	35	45	57	—
Ba^{139}	84 min	0.10	—	16	20	—	30	30	—	—
Ba^{140}	12.8 days	0.10	—	11	15	23	22	21	—	29
Ce^{141}	33.1 days	—	2.1	14	—	28	49	—	—	38
Ce^{143}	33 h	—	1.5	12	—	34	27	—	—	31
Ce^{144}	285 days	—	0.98	7.4	—	16	15	—	—	20
Nd^{147}	11.06 days	—	—	7.1	—	15	20	—	—	16
Eu^{156}	15.4 days	—	—	0.55	—	2.1	3.6	—	—	2.4
Eu^{157}	15.4 h	—	—	0.56	—	1.95	2.7	—	—	1.8
Gd^{159}	18 h	—	—	0.29	—	0.55	—	—	—	1.3
Tb^{161}	6.88 days	—	—	—	—	—	0.36	—	—	0.67

TABLE 34. Fragment Yields in the Fission of U^{238} by
Helium Nuclei (mbarn) [136]

Nuclide	Half-life	Energy, MeV								
		22.6	25.2	27.1	32.5	33.8	38.6	40	43.9	45.4
Sr^{89}	50.5 days	—	—	—	—	—	—	24	—	27
Sr^{91}	9.67 h	—	—	—	—	—	—	27	—	35
Zr^{95}	65 days	4.7	—	29	21	28	38	35	41	36
Zr^{97}	17.0 h	8.0	—	36	34	41	54	54	53	52
Mo^{99}	66 h	—	—	—	—	—	—	59	—	—
Ru^{103}	39.8 days	6.5	—	—	—	—	47	44	51	47
Ru^{105}	4.5 h	7.0	—	—	—	—	36	53	55	48
Pd^{112}	21 h	...	—	—	—	—	—	54	—	—
Ag^{111}	7.6 days	—	—	—	—	—	—	43	—	—
Ag^{113}	5.3 h	—	—	—	—	—	—	49	—	—
Cd^{115}	53 h	2.6	—	15.4	—	—	48	60	58	49
Cd^{117}	~50 min	1.9	—	—	—	—	—	—	61	—
Te^{129m}	33.5 days	—	—	—	—	—	—	31	—	—
Te^{132}	77.7 h	—	—	—	—	—	—	39	—	—
Ba^{130}	84 min	6.5	—	—	—	—	36	—	37	42
Ba^{140}	12.8 days	5.8	—	—	—	—	35	35	36	36
Ce^{141}	33.1 days	—	—	40	—	—	—	—	—	—
Ce^{143}	33 h	—	11.5	23	—	—	44	49	30	—
Nd^{147}	11.06 days	—	—	15	—	—	27	19	—	—
Eu^{156}	15.4 days	—	—	1.8	—	—	3.4	4.1	—	—
Eu^{157}	15.4 h	—	—	1.5	—	—	2.5	2.2	—	—
Gd^{159}	18 h	—	—	—	—	—	—	0.71	—	—
Tb^{161}	6.88 days	—	—	0.29	—	—	0.46	—	—	—

TABLE 35. Fragment Yields in the Fission of U^{238} by Helium Nuclei (mbarn) [137]

Nuclide	Half-life	Energy, MeV					
		19.8	24.1	31.0	33.8	36.8	39.9
Zn^{72}	49 h	—	—	—	—	—	0.54
Br^{83}	2.3 h	—	—	—	2.7±0.3	2.2	5.3±0.5
Sr^{89}	50.5 days	—	—	3.66	8.8±0.9	22.5	14±1.0
Sr^{91}	9.67 h	0.08	1.1±0.1	15.6	16.5±0.7	29.4±0.9	29±2.0
Sr^{92}	2.60 h	—	—	—	17.7±1.1	—	32.1±1.0
Y^{93}	10.4 h	--	3.5	22.0±0.2	29.7±1.6	35.6±1.4	40.4±1.3
Zr^{95}	65 days	0.28	5.8±0.1	35.1±1.5	43.5±2.8	48.6±2.6	50.5±1.8
Zr^{97}	17.0 h	0.33	6.6	37.2±1.3	42.8±1.9	47.2±0.4	49.6±2.4
Ru^{103}	39.8 days	0.35	5.48	—	42.0±0.7	43.8	49.7±3.0
Ru^{105}	4.5 h	--	4.74	27.7	31.8±1.3	37.5	37.6±1.3
Ru^{106}	1.0 yr	—	—	—	36.4±0.4	—	42.3±0.3
Pd^{112}	21 h	0.03	2.0±0.1	19.0±0.5	—	30.4±2.1	39.6
$Cd^{115,\,115m}$	53 h 43 days	0.019	1.48	16.3	22	30.4	37.3±0.8

Isotope	Half-life						
I^{131}	8.08 days	0.13	3.71±0.4	25.4	29.1±0.6	31.6±1.6	40.9±1.1
I^{133}	20.8 h	0.23	4.1±0.1	26.3±1.7	31.4±0.5	—	39.4
Ba^{139}	84 min	—	—	—	—	—	43.9
Ba^{140}	12.8 days	0.341	5.7±0.1	30.8±0.9	—	31.1±0.6	37.5±2.5
Ce^{141}	33.1 days	0.34	5.2	29.2	33.4	27.2±1.7	47.2±0.3
Ce^{143}	33 h	0.32	5.2	27.3±0.1	27.8	28.2±0.4	43.0±0.5
Pr^{145}	5.95 h	—	—	—	17.9±1.9	—	35.7±0.6
Nd^{147}	11.06 days	0.182	2.43±0.15	6.74±0.1	13.0±1.0	15.1±0.4	19.0±0.9
Sm^{153}	47.1 h	0.017	0.68	0.40	3.5±0.5	2.1±0.2	5.5±0.8
Eu^{157}	15.4 h	—	—		1.6±0.5	—	1.8±0.2
Gd^{159}	18 h	—	—		1.0±0.3	1.0	1.5±0.1

TABLE 36. Fragment Yields in the Fission of Pu[238] by Helium Nuclei (mbarn) [139]

Nuclide	Half-life	Energy, MeV						
		25.2	28.7	30.2	33.0	36.6	42.2	47.4
Sr[91]	9,67 h	—	14	—	17	12	11	29
Cd[115, 115m]	53 h, 43 days	10	23	28	43	38	43	57
Ba[140]	12.8 days	10	12	17	20	22	15	19
Ce[143]	33 h	—	—	25	—	34	22	—
Nd[147]	11.06 days	—	—	17	—	23	20	—
Eu[156]	15.4 days	—	—	2.8	—	3.6	3.7	—
Tb[161]	6.88 days	—	—	0.48	—	1.9	1.0	—

TABLE 37. Fragment Yields in the Fission of Pu[239] by Helium Nuclei (mbarn) [139]

Nuclide	Half-life	Energy, MeV									
		20.2	24.0	24.5	27.5	34	38	40.7	43.8	46	47.5
Br[82]	35.87 h	—	0.05	—	—	—	—	—	—	—	—
Br[83]	2.3 h	—	0.23	—	—	—	—	—	—	—	—
Sr[89]	50.5 days	0.06	0.95	2.0	3.4	4.3	15	9.5	16	14	15
Sr[91]	9.67 h	0.11	2.1	3.5	—	5.4	14	15	26	17	34
Sr[92]	2.6 h	0.13	1.4	—	—	6.2	13	13	—	17	21
Ru[105]	4.5 h	—	—	—	—	8.6	—	—	—	—	—
Cd[115, 115m]	53 h, 43 days	0.048	0.7	1.4	2.6	5.3	9	38	68	55	—
Cd[117]	~50 min	0.04	—	—	—	6.2	18	23	—	48	85
I[131]	8.08 days	—	1.6	—	—	—	—	—	—	—	—
I[133]	20.8 h	—	2.1	—	—	—	—	—	—	—	—
Ba[139]	84 min	0.29	—	5.4	5.7	—	—	16	—	20	21
Ba[140]	12.8 days	0.19	3.1	4.5	6.3	12	16	9.5	19	18	18
Ce[143]	33 h	—	3.0	—	—	11	—	—	—	—	—
Ce[145]	3.0 min	—	—	—	—	14	—	—	—	—	—
Nd[147]	11.06 days	—	—	—	—	—	—	—	—	12	—
Eu[156]	15.4 days	—	0.32	—	—	0.7	—	—	—	2.9	—
Eu[157]	15.4 h	—	0.30	—	—	—	—	—	—	2.3	—
Tb[161]	6.88 days	—	—	—	—	—	—	—	—	1.9	—

100

TABLE 38. Yields of Br[82], I[130], I[135], La[140], Pr[142], and Their Ratios to the Total Yields of the Respective Chains (Y_i/Y_{total}) in the Fission of U[233], U[235], U[238] by Helium Nuclei of Energies in the 25.3 to 40.5 MeV Range [148]

	Nuclide	Half-life	U²³³				U²³⁵			U²³⁸	
			25.3 MeV	29 MeV	34.5 MeV	40.5 MeV	28.2 MeV	33.8 MeV	39.9 MeV	33.8 MeV	39.9 MeV
Yield, mbarn	Br⁸²	35.87	0.106	0.20	0.45	0.59	0.064	0.12	0.20	0.038	0.075
	I¹³⁰	12.6	4.93	8.6	16.0	22.1	4.5	11.1	17.1	3.78	7.0
	I¹³⁵	6.68	2.3	—	4.4	—	—	—	—	—	—
	La¹⁴⁰	40.22	3.3	6.2	8.9	10.9	2.7	4.3	8.5	1.8	4.2
	Pr¹⁴²	19.2	0.23	0.38	0.63	1.41	0.22	0.48	0.67	—	—
$\dfrac{Y_i}{Y_{total}}$, %	Br⁸²	35.87	3.8	4.2	6.4	9.09	2.3	2.5	2.86	1.0	1.56
	I¹³⁰	12.6	36.1	39	39.1	45.1	20.5	30	35.2	10.2	15.8
	I¹³⁵	6.68	12.8	—	9.0	—	—	—	—	—	—
	La¹⁴⁰	40.22	20.1	26.9	26.2	31	11.7	12.6	21	4.2	8.4
	Pr¹⁴²	19.2	1.9	2.1	2.6	6.1	1.0	1.45	1.9	—	—

TABLE 39. Fragment Yields in the Fission of Gold by C^{12} Ions of Energies 76, 95, and 112 MeV, in mbarn [155]

Nuclide	Half-life	C^{12} ion energy, MeV		
		112	95	76
Ni^{65}	2.56 h	1.36 ± 0.14	—	—
Ni^{66}	54.6 h	1.25 ± 0.13	—	—
As^{74}	17.5 days	0.47 ± 0.10	—	—
As^{76}	26.4 h	$\begin{cases} 2.8 \pm 1.4 \\ 3.7 \pm 0.8 \end{cases}$	—	—
As^{77}	38.7 h	10.0 ± 2	—	—
As^{78}	90 min	5.6 ± 1	—	—
Br^{80}	18 min	~ 0	—	—
Br^{80m}	4.4 h	5.4 ± 1.1	—	—
Br^{82}	35.9 h	12 ± 2.4	—	—
Br^{83}	2.3 h	17.5 ± 3.5	. . .	..
Br^{84m}	6.0 min	5.5 ± 1.1	—	—
Br^{84}	32 min	2 ± 0.4	—	—
Sr^{89}	50.5 days	38.8 ± 8.8	—	—
Sr^{90}	27.7 yr	33 ± 5	—	—
Sr^{91}	9.67 h	20.4 ± 3	—	—
Y^{90}	64.2 h	16.3 ± 3	8.9 ± 0.9	1.05 ± 0.1
Y^{91}	57.5 days	24 ± 5	17.5 ± 2	3.00 ± 0.3
Y^{92}	3.60 h	25.7 ± 5	23 ± 2	5.30 ± 0.5
Y^{93}	10.4 h	24.5 ± 5	27 ± 3	8.7 ± 0.9
Y^{94}	16.5 min	12.8 ± 3.6	21 ± 4	8.4 ± 1.6
Y^{95}	10.5 min	4.4 ± 2.2	—	—
Zr^{95}	65 days	34 ± 7	—	—
Zr^{97}	17 h	11.6 ± 2	—	—
		11.5 ± 2	—	—
Nb^{95}	35 days	17 ± 3	9.4 ± 2	$\leqslant 1.3$
Nb^{96}	23.3 h	29 ± 6	16.0 ± 3	9.7
Nb^{97}	72.1 min	38 ± 8	32 ± 6	18

TABLE 39 (Continued)

Nuclide	Half-life	C^{12} ion energy, MeV		
		112	95	76
Nb^{98}	51.5 min	24 ± 6	15 ± 3	19
Mo^{99}	66 h	45 ± 10	—	—
Ag^{111}	7.6 days	17 ± 3	—	—
Ag^{112}	3.20 h	10.8 ± 2	—	—
Ag^{113}	5.3 h	7.6 ± 1.5	—	—
Ag^{115}	21.1 min	0.92 ± 0.20	—	—
I^{121}	2.0 h	0.04 ± 0.01	—	—
I^{123}	13.0 h	0.44 ± 0.12	—	—
I^{124}	4.2 days	1.20 ± 0.30	—	—
I^{125}	60 days	3.2 ± 0.6	—	—
I^{126}	13.3 days	3.1 ± 0.6	—	—
		3.1		
		3.2		
I^{128}	25.0 min	1.0 ± 0.3	—	—
I^{130}	12.6 h	$\geqslant 0.077 \pm 0.02$	—	—
		$\geqslant 0.087$		
		$\geqslant 0.100$		
Cs^{127}	6.3 h	0.079 ± 0.02	—	—
Cs^{129}	30.7 h	0.6 ± 0.12	—	—
Cs^{131}	9.6 days	1.1 ± 0.22	—	—
Cs^{132}	6.2 days	0.9 ± 0.18	·	—
Cs^{135}	$3 \cdot 10^6$ yr	0.090 ± 0.018	—	—
Ba^{131}	11.5 days	0.16 ± 0.04	--	—

5. SCHEMES OF DECAY CHAINS

Below, we list some schemes of chains of radioactive decay for cases of fission of U^{235} by thermal neutrons [6].

The underlined figures indicate values of the yields of the corresponding fragments (in percentage). These figures provide an idea of the independent, cumulative, and total yields for the several links in a particular chain.

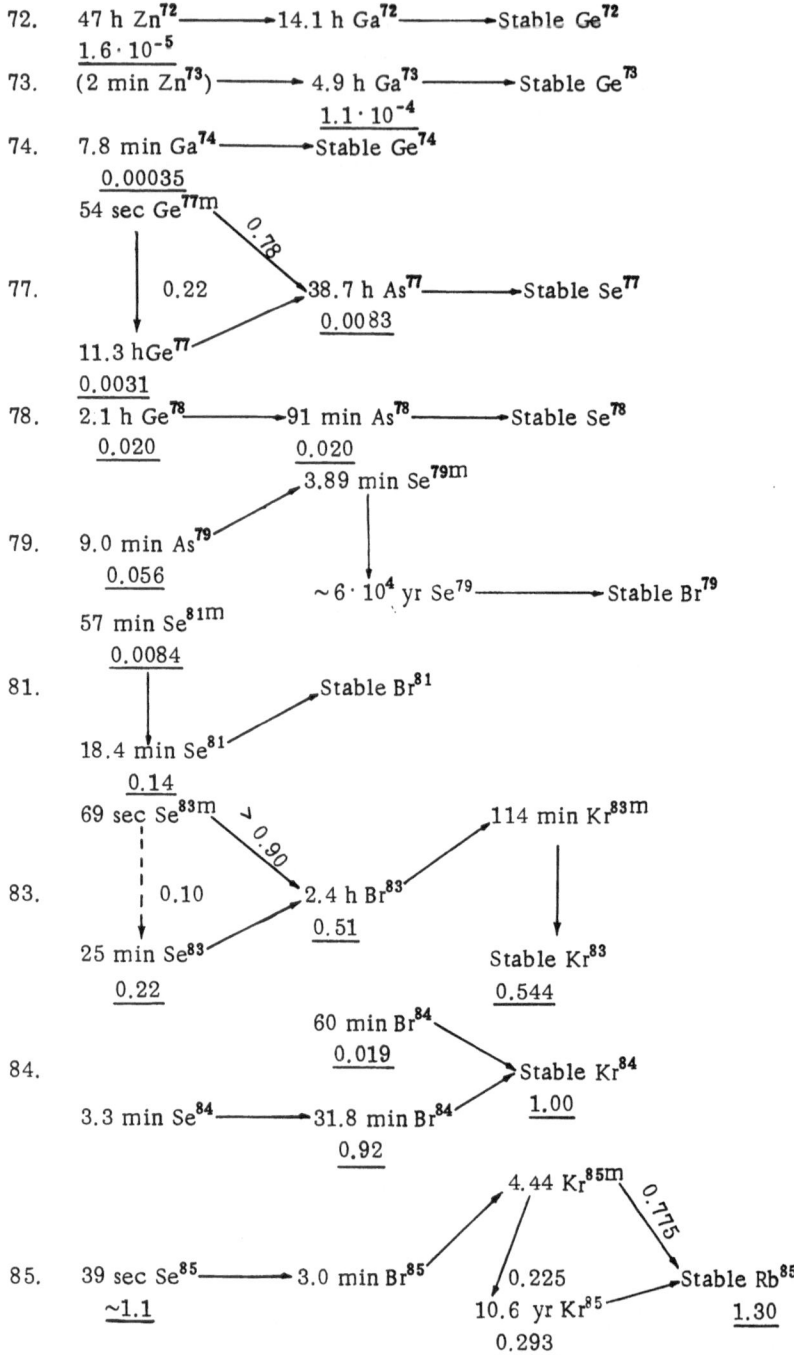

72. 47 h Zn^{72} ⟶ 14.1 h Ga^{72} ⟶ Stable Ge^{72}

$\underline{1.6 \cdot 10^{-5}}$

73. (2 min Zn^{73}) ⟶ 4.9 h Ga^{73} ⟶ Stable Ge^{73}

$\underline{1.1 \cdot 10^{-4}}$

74. 7.8 min Ga^{74} ⟶ Stable Ge^{74}

$\underline{0.00035}$

54 sec Ge^{77m}

0.78

77. 0.22 ⟶ 38.7 h As^{77} ⟶ Stable Se^{77}

$\underline{0.0083}$

11.3 hGe^{77}

$\underline{0.0031}$

78. 2.1 h Ge^{78} ⟶ 91 min As^{78} ⟶ Stable Se^{78}

$\underline{0.020}$ $\underline{0.020}$

3.89 min Se^{79m}

79. 9.0 min As^{79}

$\underline{0.056}$

~6 · 10^4 yr Se^{79} ⟶ Stable Br^{79}

57 min Se^{81m}

$\underline{0.0084}$

81. ⟶ Stable Br^{81}

18.4 min Se^{81}

$\underline{0.14}$

69 sec Se^{83m}

0.90

114 min Kr^{83m}

83. 0.10 ⟶ 2.4 h Br^{83}

$\underline{0.51}$

25 min Se^{83}

$\underline{0.22}$

Stable Kr^{83}

$\underline{0.544}$

60 min Br^{84}

$\underline{0.019}$

84. ⟶ Stable Kr^{84}

$\underline{1.00}$

3.3 min Se^{84} ⟶ 31.8 min Br^{84}

$\underline{0.92}$

4.44 Kr^{85m}

0.775

85. 39 sec Se^{85} ⟶ 3.0 min Br^{85} ⟶ Stable Rb^{85}

$\underline{\sim 1.1}$ 0.225 $\underline{1.30}$

10.6 yr Kr^{85}

$\underline{0.293}$

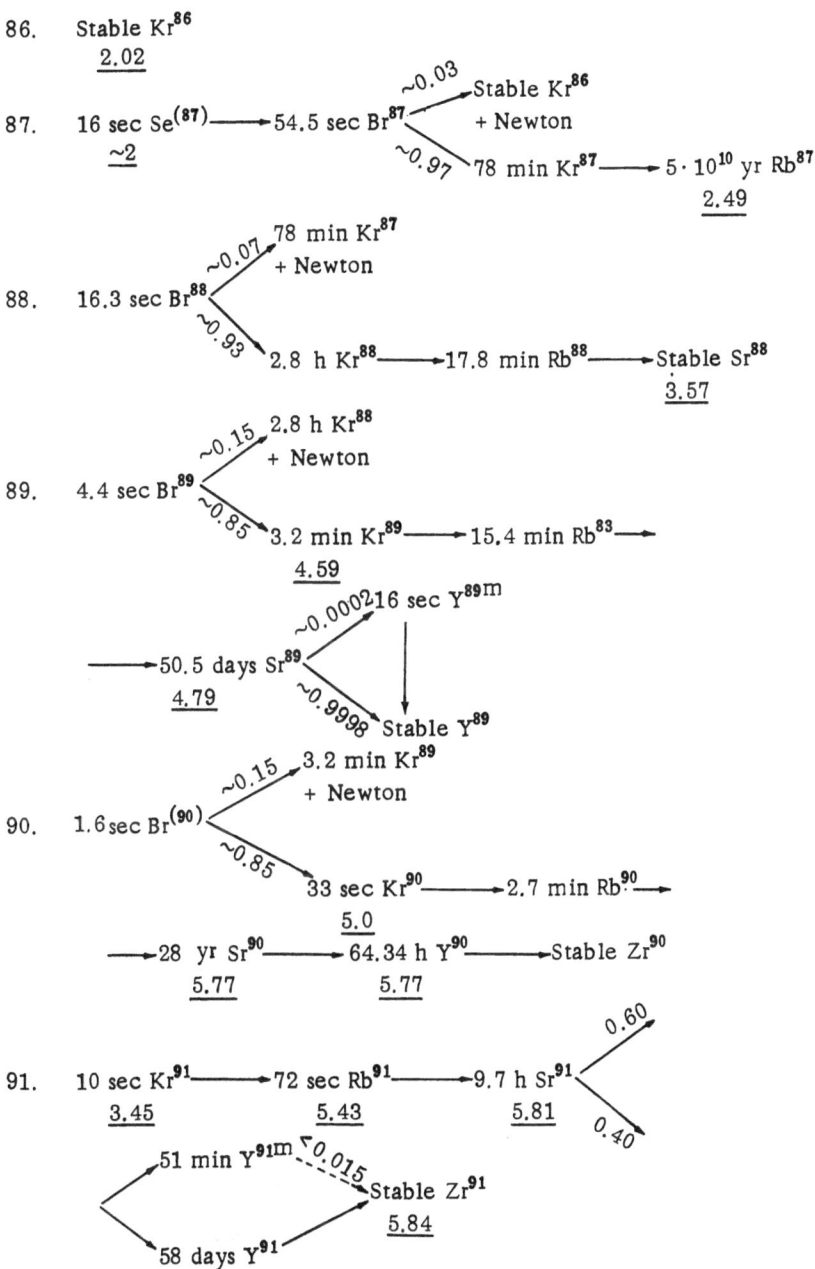

86. Stable Kr^{86}
    2.02

87. 16 sec Se$^{(87)}$ ──► 54.5 sec Br^{87} ~0.03 ──► Stable Kr^{86}
    ~2   + Newton
                        ~0.97 ╲ 78 min Kr^{87} ──► 5·10^{10} yr Rb^{87}
                                                              2.49

88. 16.3 sec Br^{88} ~0.07 ──► 78 min Kr^{87}
                                 + Newton
                        ~0.93 ╲ 2.8 h Kr^{88} ──► 17.8 min Rb^{88} ──► Stable Sr^{88}
                                                                                3.57

89. 4.4 sec Br^{89} ~0.15 ──► 2.8 h Kr^{88}
                                + Newton
                        ~0.85 ╲ 3.2 min Kr^{89} ──► 15.4 min Rb^{83} ──►
                                4.59

    ──► 50.5 days Sr^{89} ~0.0002 ──► 16 sec Y^{89m}
        4.79
                        ~0.9998 ╲ Stable Y^{89}

90. 1.6 sec Br$^{(90)}$ ~0.15 ──► 3.2 min Kr^{89}
                                  + Newton
                        ~0.85 ╲ 33 sec Kr^{90} ──► 2.7 min Rb^{90} ──►
                                5.0

    ──► 28 yr Sr^{90} ──► 64.34 h Y^{90} ──► Stable Zr^{90}
        5.77   5.77

91. 10 sec Kr^{91} ──► 72 sec Rb^{91} ──► 9.7 h Sr^{91} ╱ 0.60
    3.45   5.43   5.81   ╲ 0.40

    51 min Y^{91m} ◄ 0.015
                        ╲ ──► Stable Zr^{91}
    58 days Y^{91} 5.84
    ~5.4

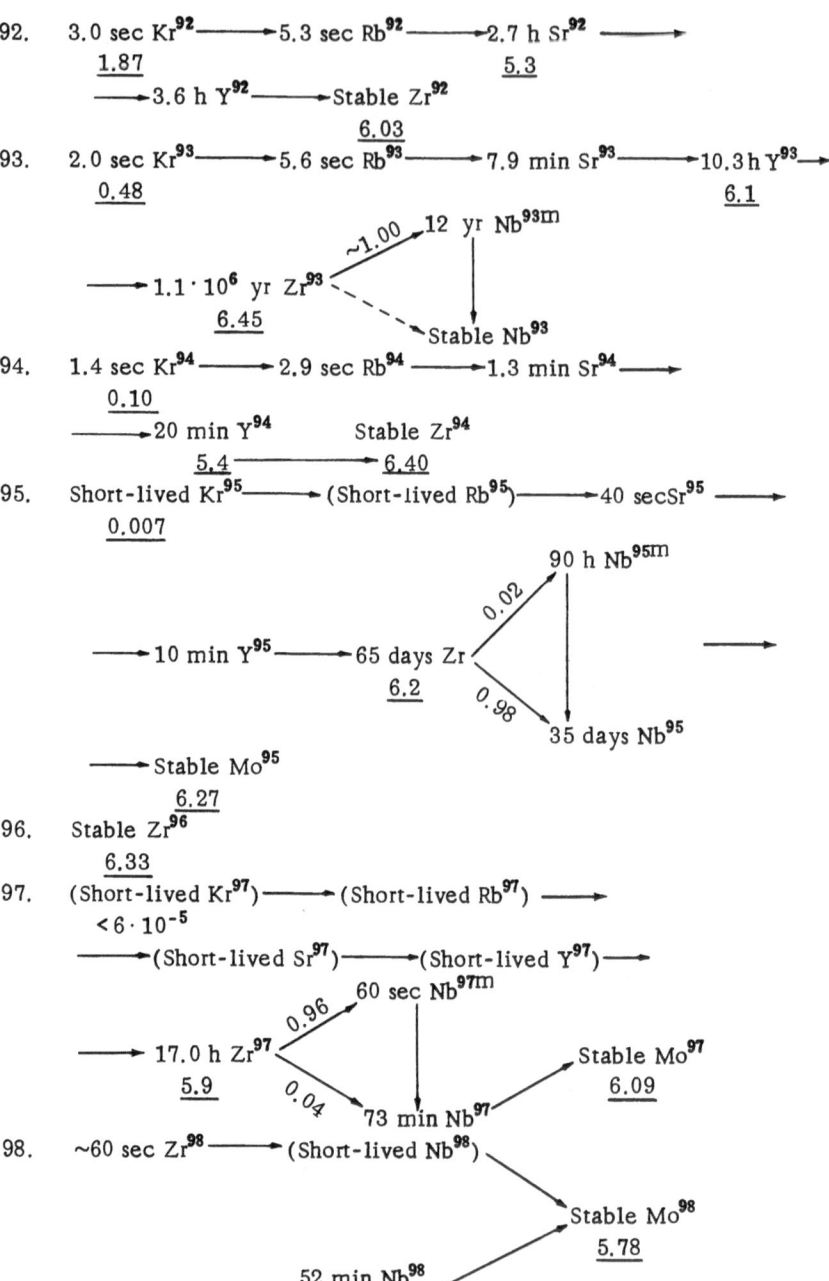

92. 3.0 sec Kr^{92} ⟶ 5.3 sec Rb^{92} ⟶ 2.7 h Sr^{92} ⟶
     <u>1.87</u>   <u>5.3</u>
     ⟶ 3.6 h Y^{92} ⟶ Stable Zr^{92}
                              <u>6.03</u>

93. 2.0 sec Kr^{93} ⟶ 5.6 sec Rb^{93} ⟶ 7.9 min Sr^{93} ⟶ 10.3 h Y^{93} ⟶
     <u>0.48</u>   <u>6.1</u>

                                    ~1.00 12 yr Nb^{93m}

     ⟶ 1.1·10^6 yr Zr^{93}
              <u>6.45</u>
                                         Stable Nb^{93}

94. 1.4 sec Kr^{94} ⟶ 2.9 sec Rb^{94} ⟶ 1.3 min Sr^{94} ⟶
     <u>0.10</u>
     ⟶ 20 min Y^{94} Stable Zr^{94}
          <u>5.4</u> ⟶ <u>6.40</u>

95. Short-lived Kr^{95} ⟶ (Short-lived Rb^{95}) ⟶ 40 sec Sr^{95} ⟶
     <u>0.007</u>

                                               90 h Nb^{95m}

                                    0.02

     ⟶ 10 min Y^{95} ⟶ 65 days Zr   ⟶
                              <u>6.2</u>
                                    0.98

                                               35 days Nb^{95}

     ⟶ Stable Mo^{95}
          <u>6.27</u>

96. Stable Zr^{96}
     <u>6.33</u>

97. (Short-lived Kr^{97}) ⟶ (Short-lived Rb^{97}) ⟶
     < 6·10^{-5}
     ⟶ (Short-lived Sr^{97}) ⟶ (Short-lived Y^{97}) ⟶

                                    60 sec Nb^{97m}
                              0.96
                                                       Stable Mo^{97}
     ⟶ 17.0 h Zr^{97} <u>6.09</u>
          <u>5.9</u>   0.04
                              73 min Nb^{97}

98. ~60 sec Zr^{98} ⟶ (Short-lived Nb^{98})

                                                       Stable Mo^{98}
                                                            <u>5.78</u>

                    52 min Nb^{98}
                         <u>0.064</u>

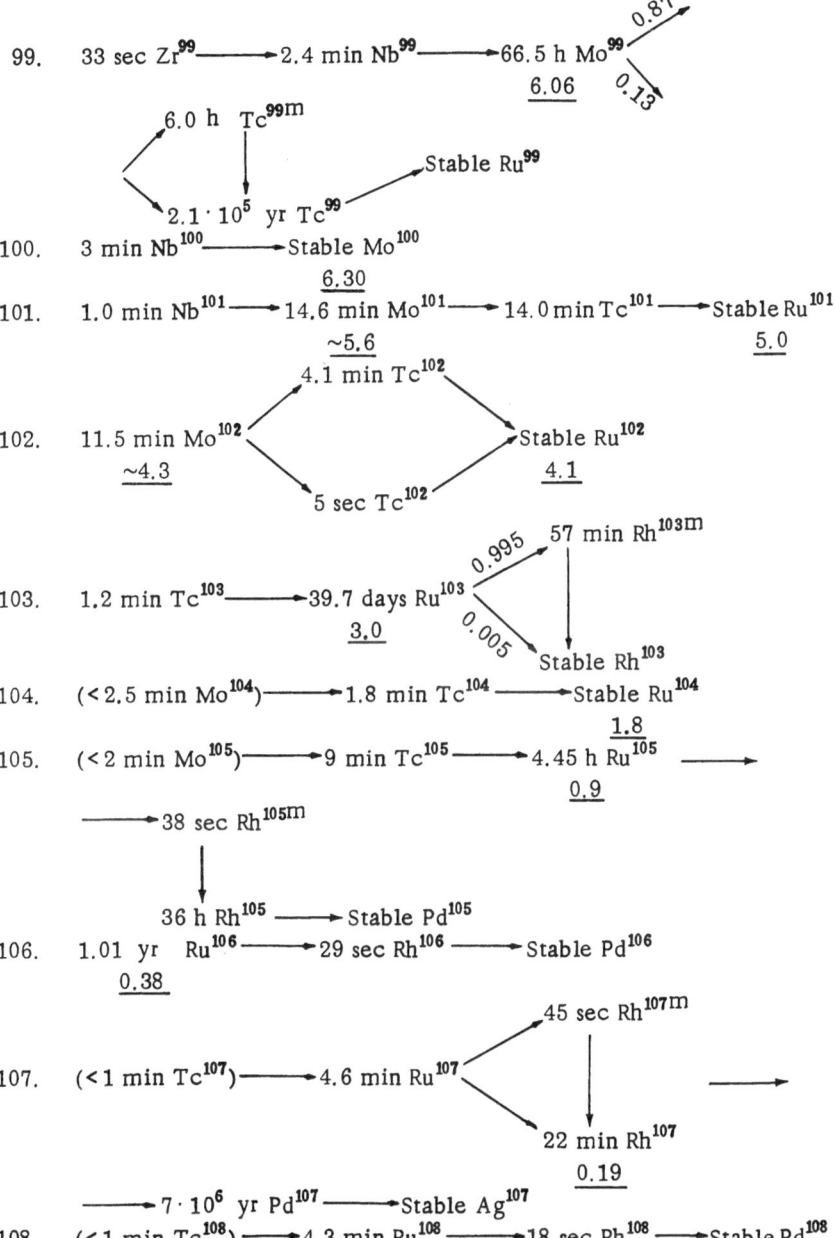

99.   33 sec Zr⁹⁹ ——→ 2.4 min Nb⁹⁹ ——→ 66.5 h Mo⁹⁹

$$33 \text{ sec } Zr^{99} \longrightarrow 2.4 \text{ min } Nb^{99} \longrightarrow 66.5 \text{ h } Mo^{99}$$

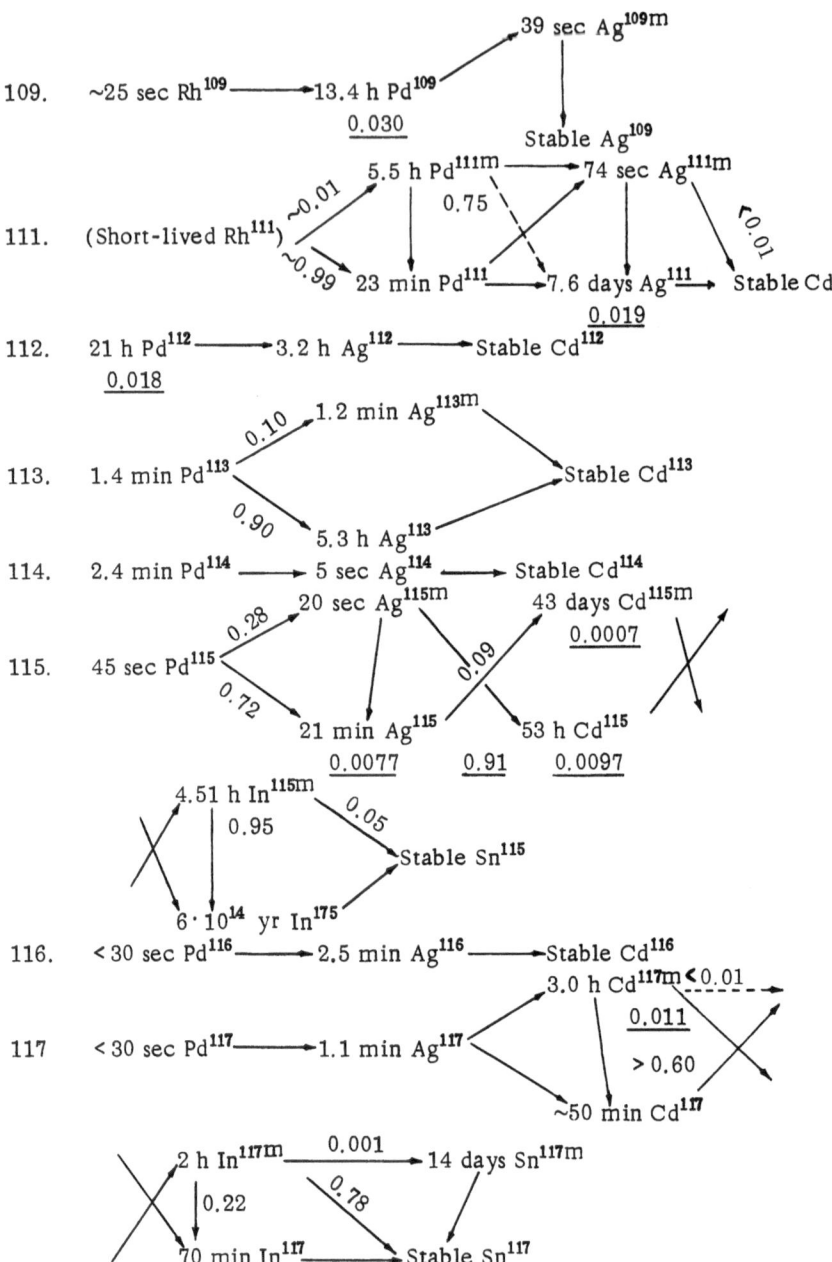

109.   ~25 sec Rh^{109} ────→ 13.4 h Pd^{109} → 39 sec Ag^{109m}

                                      0.030   Stable Ag^{109}

111.   (Short-lived Rh^{111})   ~0.01 ↗ 5.5 h Pd^{111m} ────→ 74 sec Ag^{111m}

                                 0.75   <0.01

                        ~0.99 ↘ 23 min Pd^{111} ──→ 7.6 days Ag^{111} → Stable Cd

                                                  0.019

112.   21 h Pd^{112} ────→ 3.2 h Ag^{112} ────→ Stable Cd^{112}

     0.018

113.   1.4 min Pd^{113} 0.10 ↗ 1.2 min Ag^{113m} ↘ Stable Cd^{113}

                          0.90 ↘ 5.3 h Ag^{113} ↗

114.   2.4 min Pd^{114} ────→ 5 sec Ag^{114} ────→ Stable Cd^{114}

115.   45 sec Pd^{115} 0.28 ↗ 20 sec Ag^{115m} 43 days Cd^{115m}

                                          0.0007

                0.72 ↘ 21 min Ag^{115} 0.09   53 h Cd^{115}

                               0.0077   0.91   0.0097

     4.51 h In^{115m} 0.05

               0.95   Stable Sn^{115}

     6·10^{14} yr In^{175}

116.   <30 sec Pd^{116} ────→ 2.5 min Ag^{116} ────→ Stable Cd^{116}

                                        3.0 h Cd^{117m} <0.01

117   <30 sec Pd^{117} ────→ 1.1 min Ag^{117} 0.011

                                           >0.60

                                  ~50 min Cd^{117}

     2 h In^{117m} 0.001 ────→ 14 days Sn^{117m}

              0.22   0.78

     70 min In^{117} ────────→ Stable Sn^{117}

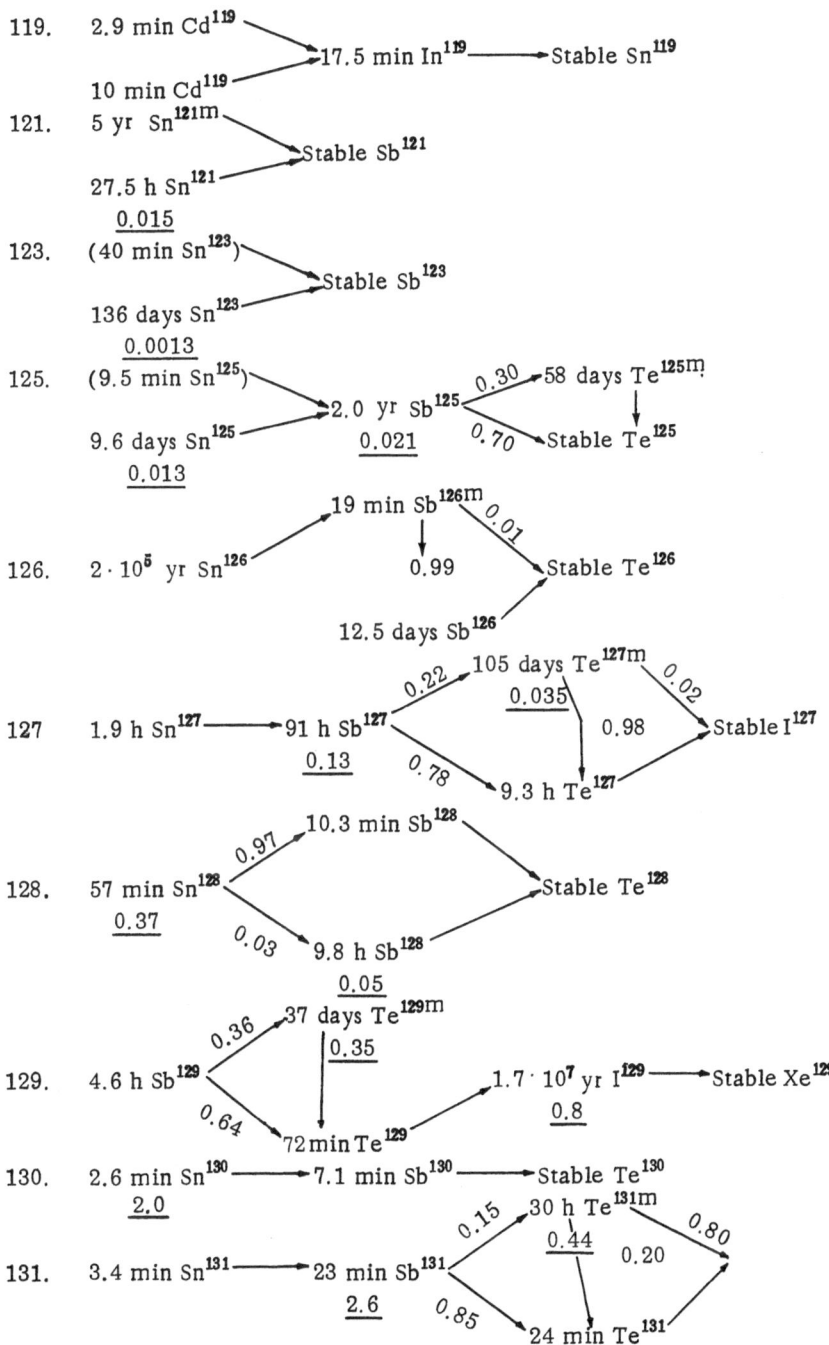

119. 2.9 min Cd^{119}

10 min Cd^{119} → 17.5 min In^{119} → Stable Sn^{119}

121. 5 yr Sn^{121m}

27.5 h Sn^{121} → Stable Sb^{121}

123. $\underline{0.015}$ (40 min Sn^{123})

136 days Sn^{123} → Stable Sb^{123}

125. $\underline{0.0013}$ (9.5 min Sn^{125})

9.6 days Sn^{125} → 2.0 yr Sb^{125} $\underline{0.021}$ → 0.30 → 58 days Te^{125m} → 0.70 → Stable Te^{125}

$\underline{0.013}$

126. 2·10^5 yr Sn^{126} → 19 min Sb^{126m} → 0.99, 0.01 → Stable Te^{126}

12.5 days Sb^{126}

127. 1.9 h Sn^{127} → 91 h Sb^{127} $\underline{0.13}$ → 0.22 → 105 days Te^{127m} $\underline{0.035}$ → 0.98, 0.02 → 0.78 → 9.3 h Te^{127} → Stable I^{127}

128. 57 min Sn^{128} $\underline{0.37}$ → 0.97 → 10.3 min Sb^{128} → 0.03 → 9.8 h Sb^{128} → Stable Te^{128}

$\underline{0.05}$

129. 4.6 h Sb^{129} → 0.36 → 37 days Te^{129m} $\underline{0.35}$ → 0.64 → 72 min Te^{129} → 1.7·10^7 yr I^{129} $\underline{0.8}$ → Stable Xe^{129}

130. 2.6 min Sn^{130} $\underline{2.0}$ → 7.1 min Sb^{130} → Stable Te^{130}

131. 3.4 min Sn^{131} → 23 min Sb^{131} $\underline{2.6}$ → 0.15 → 30 h Te^{131m} $\underline{0.44}$ → 0.20, 0.80 → 0.85 → 24 min Te^{131}

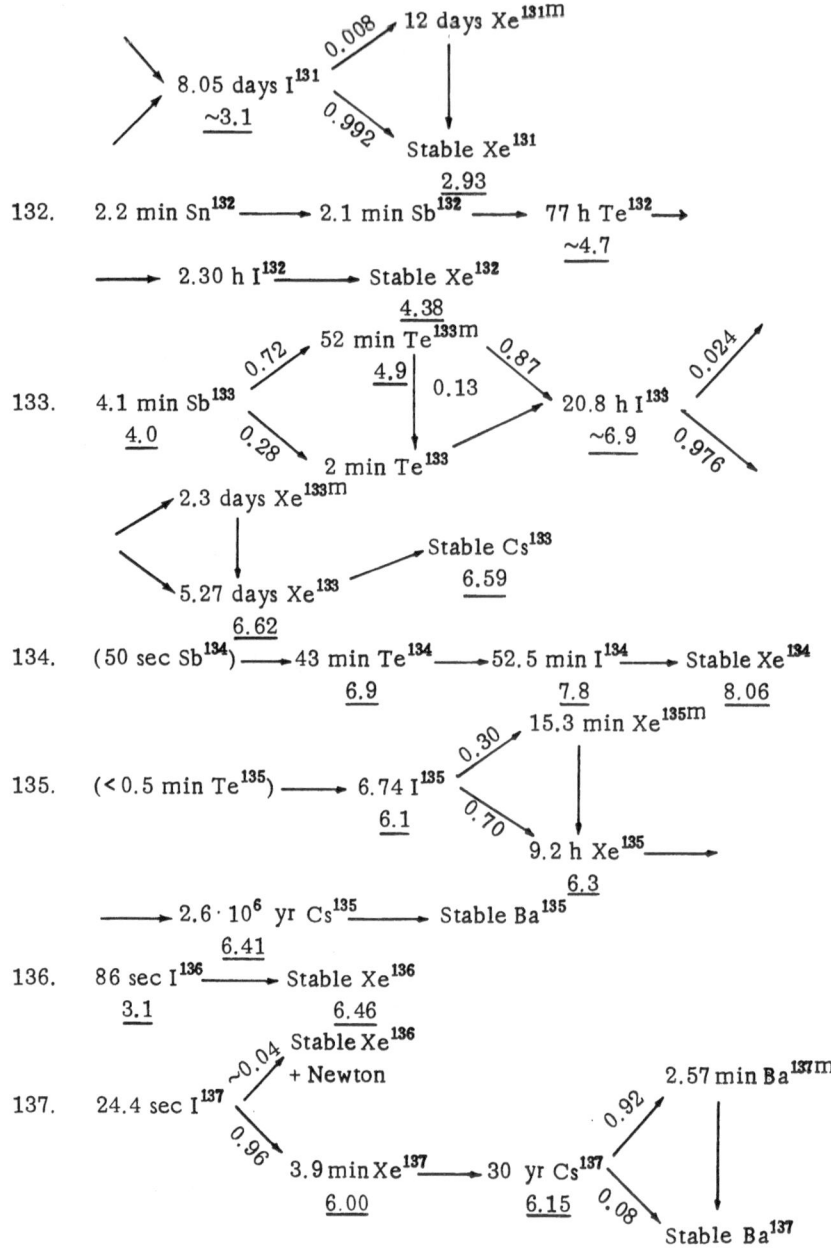

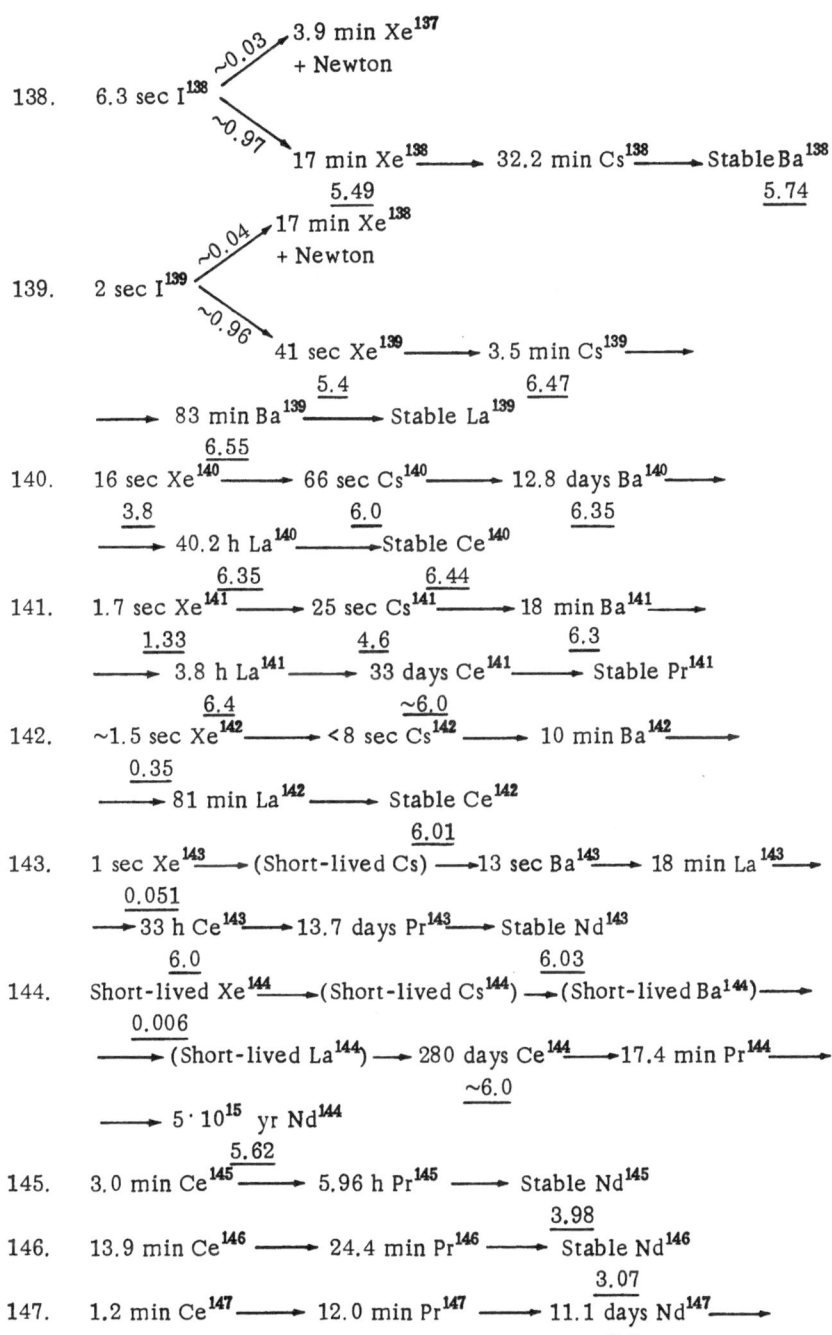

138. 6.3 sec I^{138} →~0.03→ 3.9 min Xe^{137} + Newton

→~0.97→ 17 min Xe^{138} → 32.2 min Cs^{138} → Stable Ba^{138}
$\underline{5.49}$ $\underline{5.74}$

139. 2 sec I^{139} →~0.04→ 17 min Xe^{138} + Newton

→~0.96→ 41 sec Xe^{139} → 3.5 min Cs^{139} →
$\underline{5.4}$ $\underline{6.47}$

→ 83 min Ba^{139} → Stable La^{139}
$\underline{6.55}$

140. 16 sec Xe^{140} → 66 sec Cs^{140} → 12.8 days Ba^{140} →
$\underline{3.8}$ $\underline{6.0}$ $\underline{6.35}$

→ 40.2 h La^{140} → Stable Ce^{140}
$\underline{6.35}$ $\underline{6.44}$

141. 1.7 sec Xe^{141} → 25 sec Cs^{141} → 18 min Ba^{141} →
$\underline{1.33}$ $\underline{4.6}$ $\underline{6.3}$

→ 3.8 h La^{141} → 33 days Ce^{141} → Stable Pr^{141}
$\underline{6.4}$ $\underline{\sim 6.0}$

142. ~1.5 sec Xe^{142} → <8 sec Cs^{142} → 10 min Ba^{142} →
$\underline{0.35}$

→ 81 min La^{142} → Stable Ce^{142}
$\underline{6.01}$

143. 1 sec Xe^{143} → (Short-lived Cs) → 13 sec Ba^{143} → 18 min La^{143} →
$\underline{0.051}$

→ 33 h Ce^{143} → 13.7 days Pr^{143} → Stable Nd^{143}
$\underline{6.0}$ $\underline{6.03}$

144. Short-lived Xe^{144} → (Short-lived Cs^{144}) → (Short-lived Ba^{144}) →
$\underline{0.006}$

→ (Short-lived La^{144}) → 280 days Ce^{144} → 17.4 min Pr^{144} →
$\underline{\sim 6.0}$

→ 5·10^{15} yr Nd^{144}
$\underline{5.62}$

145. 3.0 min Ce^{145} → 5.96 h Pr^{145} → Stable Nd^{145}
$\underline{3.98}$

146. 13.9 min Ce^{146} → 24.4 min Pr^{146} → Stable Nd^{146}
$\underline{3.07}$

147. 1.2 min Ce^{147} → 12.0 min Pr^{147} → 11.1 days Nd^{147} →
$\underline{\sim 2.7}$

$\longrightarrow$ 2.6 yr Pm^{147} $\longrightarrow$ 1.3·10^{11} yr Sm^{147}

148.   40 sec Ce^{148} $\longrightarrow$ 1.95 min Pr^{148} $\xrightarrow{\underline{2.36}}$ Stable Nd^{148}

149.   (2.0 h Nd^{149}) $\longrightarrow$ 53.1 h Pm^{149} $\xrightarrow{\underline{1.71}}$ Stable Sm^{149}

150.   Stable Nd^{150} $\underline{1.13}$

151.   (13 $\overline{\text{min}}$ Nd^{151}) $\xrightarrow{0.67}$ 28.4 h Pm^{151} $\longrightarrow$ 80 yr Sm^{151} $\longrightarrow$ Stable Eu^{151}
       $\underline{0.44}$

152.   Stable Sm^{152}

153.   47 h Sm^{153} $\xrightarrow{0.281}$ Stable Eu^{153}

154.   Stable Sm^{154} $\underline{0.15}$ $\underline{0.169}$

155.   24 min Sm^{155} $\xrightarrow{\underline{0.077}}$ 4 yr Ed^{155} $\longrightarrow$ Stable Gd^{155}

156.   9 h Sm^{156} $\xrightarrow{\underline{0.033}}$ 15.4 days Eu^{156} $\xrightarrow{\underline{0.033}}$ Stable Gd^{156}

157.   15.4 h Eu^{157} $\xrightarrow{\underline{0.013}}$ Stable Gd^{157} $\underline{0.014}$

158.   60 min Eu^{158} $\xrightarrow{\underline{0.0078}}$ Stable Gd^{158}

159.   18.0 h Gd^{159} $\xrightarrow{\underline{0.002}}$ Stable Tb^{159}

161.   (3.7 min Gd^{181}) $\xrightarrow{\underline{0.00107}}$ 6.9 days Tb^{161} $\longrightarrow$ Stable Dy^{161}

166.   82 h Dy^{166} $\xrightarrow{\underline{7.6·10^{-5}}}$ 27.3 h Mo^{166} $\longrightarrow$ Stable Er^{166}

BASIC REFERENCE LITERATURE

1. S. Katcoff, Nucleonics 16 (4); 78 (1958).
2. N. Gusev, V. Mashkovich, and G. Obvintsev, Gamma Emission of Radioactive Isotopes and Fission Products, Moscow, Phys.-Math. Press (1958).
3. C. D.Coryell and N. Sugarman (eds.), Radiological Studies: The Fission Products, Division 4, Vol. 9, McGraw-Hill Book Co., Inc., New York (1951).
4. E. P. Steinberg and L. E. Glendenin, Geneva 1955 conference.*
5. R. B. Duffield, R. A. Schmitt, and R. A. Sharp, Proc. Second United Nations Inter. Conf. on the Peaceful Uses of Atomic Energy 15: 202, P/678, Geneva (1958).
6. S. Katcoff, Nucleonics 18 (11): 201 (1960).
7. N. A. Perfilov, O. V. Lozhkin, and V. P. Shamov, Usp. Fiz. Nauk 70 (1): 3 (1960).
8. V. K. Gorshkov, et al., At. Energ. 3 (7): 11 (1957).
9. R. N. Ivanov, et al., At. Energ. 3 (12): 546 (1957).
10. V. K. Gorshkov, Pribory i Tekhn. Eksperim. No. 2: 53 (1957).
11. M. G. Inghram, J. Phys. Chem. 57: 809 (1953).
12. A. Dempster, Am. Phil. Soc. 75: 555 (1935).
13. V. K. Gorshkov, Physics and Heat Engineering of Reactors, Supplement No. 1 to At. Energ. (1958), Moscow, Atom Press (1958).
14. M. P. Anikina, et al., Geneva 1958 conference.
15. M. P. Anikina, et al., At. Energ. 4 (2): 198 (1958).
16. W. Fleming, R. H. Tomlinson, and H. G. Thode, Can. J. Phys. 32: 522 (1954).
17. E. P. Steinberg, et al., Phys. Rev. 95: 867 (1954).
18. J. Koch, et al., Phys. Rev. 76: 279 (1949).
19. H. G. Thode, Nucleonics 3 (3): 14 (1948).
20. D. R. Bidinosti, H. R. Fickel, and R. H. Tomlinson, Proc. of the Second United Nations Inter. Conf. on the Peaceful Uses of Atomic Energy 15: 459, P/201, Geneva (1958).

---

* References are to the conferences on the peaceful uses of atomic energy.

21. V. A. Vlasov, Yu. A. Zysin, I. S. Kirin, A. A. Lbov, L. I. Osyaeva, and L. I. Sel'chenkov, Seminar "Neutron Physics," Moscow, Atom Press, 1961, p. 235.

22. L. W. Roeland, L. W. Bollinger, and G. E. Thomas, Proc. of the Second United Nations Inter. Conf. on Peaceful Uses of Atomic Energy 15; 440, P/551, Geneva, 1958.

23. T. A. Mostovaya, Geneva 1958 conference.

24. W. Stein, Phys. Rev. 108; 94 (1957).

25. N. A. Perfilov, Physics of Fission of Atomic Nuclei, Supplement No. 1 to At. Energ. (1957) Moscow, Atom Press, 1957, p. 98.

26. J. C. Roy, Can. J. Phys. 39; 315 (1961).

27. R. B. Leachman, Geneva 1958 conference.

28. A. N. Protopopov, Seminar "Physics of the Fission of Atomic Nuclei," Moscow, Atom Press, 1962, p. 24.

29. I. P. Selinov, Geneva 1958 conference.

30. Yu. A. Zysin, A. A. Lbov, and L. I. Sel'chenkov, At. Energ. 8(5): 409 (1960).

31. E. K. Bonyushkin, et al., At. Energ. 10 (1): 13 (1961).

32. J. Halpern, Ann. Rev. Nucl. Sci. 9; 245 (1959); see also J. Halpern, Fission of Nuclei [Russian translation], Moscow, Phys.-Math. Press, 1962.

33. W. J. Swiatecki, Phys. Rev. 100; 936 (1955).

34. H. G. Thode and R. L. Graham, Can. J. Research 25A; 1 (1947).

35. J. Macnamara, C. B. Collins, and H. G. Thode, Phys. Rev. 78; 129 (1950).

36. C. W. Stanley and S. Katcoff, J. Chem. Phys. 17; 653 (1949).

37. D. Wiles and C. Coryell, Phys. Rev. 96; 696 (1954).

38. D. Wiles, et al., Can. J. Phys. 31: 419 (1953).

39. E. P. Steinberg and L. E. Glendenin, Phys. Rev. 95; 431 (1954).

40. A. W. Fairhall, R. C. Jensen, and E. F. Neuzil, Proc. of the Second United Nations Inter. Conf. on the Peaceful Uses of Atomic Energy 15: 452,P/677, Geneva, 1958.

41. R. C. Jensen and A. W. Fairhall, Phys. Rev. 109: 942 (1958).

42. N. A. Perfilov, Seminar "Physics of the Fission of Atomic Nuclei," Atom Press, 1962, p. 175.

43. V. P. Eismont, Zh. Eksperim. i Teor. Fiz. 42 (1): 178 (1962).

44. E. K. Bonyushkin, et al., Seminar "Neutron Physics," Moscow, Atom Press, 1961, p. 224.

45. W. H. Jones, et al., Phys. Rev. 99; 184 (1955).

46. J. P. Butler, Proc. of the Second United Nations Inter. Conf. on the Peaceful Uses of Atomic Energy 15; 156, Geneva, 1958.

47. J. A. Wheeler, Physica 22: 1103 (1956).

48. A. Bohr, Proc. of the Second United Nations Inter. Conf. on the Peaceful Uses of Atomic Energy 2: 151, Geneva, 1958.

49. R. B. Regier, W. H. Burgus, J. R. Smith, and M. S. Moore, Bull. Am. Phys. Soc., Ser. 2, 3: 6 (1958).

50. R. B. Regier, W. H. Burgus, and R. L. Tromp, Phys. Rev. 113: 1589 (1959).

51. R. Nasuhoglu, et al., Phys. Rev. 108: 1522 (1957).

52. Los Alamos Radiochemistry Group, Phys. Rev. 107: 325 (1957).

53. R. B. Regier, et al., Phys. Rev. 119: 2017 (1960).

54. N. I. Borisova, et al., Seminar "Physics of the Fission of Atomic Nuclei," Moscow, Atom Press, 1962, p. 48.

55. B. T. Geilikman, Seminar "Physics of the Fission of Atomic Nuclei," Moscow, Atom Press, 1962, p. 5.

56. Bohr and Wheeler, Phys. Rev. 55: 426 (1939).

57. S. Frankel and N. Metropolis, Phys. Rev. 72: 914 (1947).

58. J. Frenkel, J. Phys. USSR 1: 125 (1939).

59. R. D. Present and J. K. Knipp, Phys. Rev. 51: 75, 1188 (1940).

60. R. D. Present, F. Reines, and J. K. Knipp, Phys. Rev. 70: 557 (1946).

61. J. Frenkel, J. Phys. USSR 10: 533 (1946).

62. T. Yasaki and O. Miyatake, Phys. Rev. 79: 740 (1950); 80: 754 (1950).

63. J. Jungerman, Phys. Rev. 80: 285 (1950).

64. W. J. Swiatecki, Phys. Rev. 83: 178 (1950).

65. D. L. Hill, Bull. Am. Phys. Soc. 26: 45 (1951).

66. D. L. Hill and J. A. Wheeler, Phys. Rev. 89: 1102 (1953).

67. U. L. Businaro and S. Gallone, Nuovo cimento 1: 1277 (1955).

68. V. G. Nosov, Physics of Fission of Atomic Nuclei, Supplement No. 1 to At. Energ. (1957), Moscow, Atom Press, 1957, p. 52.

69. P. Fong, Phys. Rev. 89: 332 (1953).

70. P. Fong, Phys. Rev. 102: 434 (1956).

71. J. K. Perring and J. S. Story, Phys. Rev. 98: 1525 (1955).

72. B. T. Geilikman, At. Energ. 6 (3): 290 (1959).

73. B. T. Geilikman, Geneva 1955 conference.

74. B. T. Geilikman, Physics of Fission of Atomic Nuclei, Supplement No. 1 to At. Energ. (1957), Moscow, Atom Press, 1957, p. 5.

75. H. G. Hicks, et al., Phys. Rev. 100: 1284 (1955).

76. H. G. Hicks and R. S. Gilbert, Phys. Rev. 100: 1286 (1955).

77. W. H. Jones, et al., Phys. Rev. 99: 184 (1955).

78. M. Lindner and R. N. Osborne, Phys. Rev. 94: 1323 (1954).

79. B. J. Bowles, F. Brown, and J. P. Butler, Phys. Rev. 107: 751 (1957).

80. W. H. Fleming and H. G. Thode, Phys. Rev. 92: 378 (1953).

81. C. W. Wetherill, Phys. Rev. 92: 907 (1953).

82. J. Macnamara and H. G. Thode, Phys. Rev. 80: 471 (1950).

83. P. L. Parker and P. K. Kuroda, J. Inorg. Nucl. Chem. 5: 153 (1958).

84. P. K. Kuroda and R. R. Edwards, J. Inorg. Nucl. Chem. 3: 345 (1957).

85. L. E. Glendenin and E. P. Steinberg, J. Inorg. Nucl. Chem. 1: 45 (1955).

86. J. G. Cuninghame, J. Inorg. Nucl. Chem. 6: 181 (1958).

87. E. A. C. Crouch and J. G. Swainbank, Proc. of the Second United Nations Inter. Conf. on the Peaceful Uses of Atomic Energy 15: 464, P/7, Geneva, 1958.

88. W. E. Grummit and G. M. Milton, J. Inorg. Nucl. Chem. 5: 93 (1957).

89. M. P. Anikina and B. V. Ershler, At. Energ. 11 (3): 275 (1957).

90. D. C. Santry and L. Yaffe, Can. J. Chem. 38: 421 (1960).

91. G. W. Reed, Phys. Rev. 98: 1327 (1955).

92. F. Brown and L. Yaffe, Can. J. Chem. 31: 242 (1953).

93. L. R. Bunney, et al., Proc. of the Second United Nations Inter. Conf. on the Peaceful Uses of Atomic Energy 15: 449, P/643, Geneva, 1958.

94. J. Tarrel, et al., Phys. Rev. 92: 1091 (1953).

95. J. A. Petruska, E. A.Melaika, and R. H. Tomlinson, Can. J. Phys. 33: 640 (1955).

96. M. G. Inghram, R. J. Hayden, and D. C. Hess, Phys. Rev. 79: 27 (1950).

97. J. F. Croall, J. Inorg. Nucl. Chem. 16: 358 (1961).

98. L. R. Bunney, et al., Proc. of the Second United Nations Inter. Conf. on the Peaceful Uses of Atomic Energy 15: 444, P/644, Geneva, 1958.

99. K. Fritze, C. C. McMullen, and H. G. Thode, Proc. of the Second United Nations Inter. Conf. on the Peaceful Uses of Atomic Energy 15; 436, P/187, Geneva, 1958.

100. H. R. Fickel and R. H. Tomlinson, Can. J. Phys. 37: 926 (1959).

101. D. Wiles, J. A. Petruska, and R. H. Tomlinson, Can. J. Chem. 34; 227 (1956).

102. J. G. Bayly, et al., Can. J. Phys. 39: 1391 (1961).

103. J. G. Cuninghame, J. Inorg. Nucl. Chem. 4: 1 (1957).

104. W. E. Grummitt and G. M. Milton, J. Inorg. Nucl. Chem. 20: 6 (1961).

105. A. Turkevich and J. B. Niday, Phys. Rev. 84: 52 (1951).

106. T. J. Kennett and H. G. Thode, Can. J. Phys. 35: 969 (1957).

107. P. Kafalas and C. E. Crouthame, J. Inorg. Nucl. Chem. 4: 239 (1957).

108. Data compiled by present authors, 1953.

109. M. A. Bak, et al., At. Energ. 6 (5): 577 (1959).

110. K. A. Petrzhak, Seminar "Neutron Physics," Moscow, Atom Press, 1961, p. 217.

111. R. N. Keller, E. P. Steinberg, and L. E. Glendenin, Phys. Rev. 94: 969 (1954).

112. A. Turkevich, J. B. Niday, and A. Tompkins, Phys. Rev. 89: 552 (1953).

113. A. N. Protopopov, et al., At. Energ. 5 (2): 130 (1958).

114. A. C. Wahl and N. A. Bonner, Phys. Rev. 85: 570 (1952).

115. P. C. Stevenson, et al., Phys. Rev. 117: 186 (1960).

116. J. G. Cuninghame, J. Inorg. Nucl. Chem. 5: 1 (1957).

117. R. F. Coleman, B. E. Hawker, and J. L. Perkin, J. Inorg. Nucl. Chem. 14: 8 (1960).

118. A. C. Wahl, J. Inorg. Nucl. Chem. 6: 263 (1958).

119. I. T. Krisyuk, N. B. Platunova, and A. N. Protopopov, Radiokhimiya 2: 746 (1960).

120. I. T. Krisyuk, K. P. Lepnev, and N. B. Platunova, Radiokhimiya 2: 743 (1960).

121. N. Sugarman, Phys. Rev. 79: 532 (1950).

122. D. M. Hiller and D. S. Martin, Phys. Rev. 90: 581 (1953).

123. I. A. Vasil'ev and K. A. Petrzhak, Zh. Eksperim. i Teor. Fiz. 35: 1135 (1958).

124. R. A. Schmitt and N. Sugarman, Phys. Rev. 95: 1260 (1954).

125. H. G. Richter and C. D. Coryell, Phys. Rev. 95: 1550 (1954).

126. L. Katz, et al., Phys. Rev. 99: 98 (1955).

127. T. T. Sugihara, J. Roesmer, and J. W. Meadows, Phys. Rev. 121: 1179 (1961).

128. L. G. Jodra and N. Sugarman, Phys. Rev. 99: 1470 (1955).

129. P. C. Stevenson, et al., Phys. Rev. 111: 886 (1958).

130. R. L. Wolke, Phys. Rev. 120: 543 (1960).

131. A. W. Fairhall, Phys. Rev. 102: 1335 (1956).

132. R. C. Jensen and A. W. Fairhall, Phys. Rev. 118: 771 (1960).

133. J. M. Alexander and C. D. Coryell, Phys. Rev. 108: 1274 (1957).

134. T. T. Sugihara, et al., Phys. Rev. 108: 1264 (1957).

135. A. S. Newton, Phys. Rev. 75: 17 (1949).

136. R. Vandenbosch, et al., Phys. Rev. 111: 1358 (1958).

137. L. J. Colby, M. La Salle, and J. W. Cobble, Phys. Rev. 121: 1415 (1961).

138. R. Gunnink and J. W. Cobble, Phys. Rev. 115: 1247 (1959).

139. R. A. Glass, et al., Phys. Rev. 104: 434 (1956).

140. J. C. D. Milton and J. S. Fraser, Can. J. Phys. 40: 11, 1626 (1962).

141. J. D. Carrison and B. W. Roos, Nuclear Sci. and Eng. 12 (1): 115.

142. H. Farrar and R. H. Tomlinson, Can. J. Phys. 40 (8): 943 (1962).

143. P. K. Kuroda and M. P. Menon, Nuclear Sci. and Eng. 10 (1): 70 (1961).

144. H. Farrar, H. R. Fickel, and R. H. Tomlinson, Can. J. Phys. 40 (8): 1017 (1962).

145. H. Farrar and R. H. Tomlinson, Nuclear Phys. 34 (2): 367 (1962).

146. D. R. Bildinosti, D. E. Irish, and R. H. Tomlinson, Can. J. Chem. 39 (3): 628 (1961).

147.   A. C. Wahl, R. L. Ferguson, D. R. Nethaway, D. E. Trontner, and K. Wolfsberg, Phys. Rev. 126 (3): 1112 (1962).
148.   L. J. Colby and J. W. Cobble, Phys. Rev. 121 (5): 1410 (1961).
149.   J. G. Cuninghame, G. P. Kitt, and E. R. Rae, Nuclear Phys. 27: 1, 154 (1961).
150.   G. Rudstam and A. C. Pappas, Nuclear Phys. 22 (3): 468 (1961).
151.   H. B. Levy, et al., Phys. Rev. 124 (2): 544 (1961).
152.   K. M. Broom, Phys. Rev. 126 (2): 627 (1962).
153.   A. Kjelberg, H. Taniguchi, and L. Yaffe, Can. J. Chem. 39 (3): 635 (1961).
154.   M. Kaplan and C. D. Coryell, Phys. Rev. 124 (6): 1949 (1961).
155.   H. M. Blann, Phys. Rev. 123 (4): 1356 (1961).

SUPPLEMENTARY REFERENCE LITERATURE*

1. G. M. Kukavadze, M. P. Anikina, L. L. Gol'din, and B. V. Ershler, Session of the USSR Academy of Sciences on the Peaceful Uses of Atomic Energy, July 1-5, 1955, chemistry panel, Moscow, USSR Acad. Sci., 1955, p. 205.
2. N. Sugarman, Phys. Rev. 89: 570 (1953).
3. J. G. Cuninghame, Phil. Mag. 44: 900 (1953).
4. J. R. Arnold and N. Sugarman, J. Chem. Phys. 15: 703 (1947).
5. J. A. Petruska, H. G. Thode, and R. H. Tomlinson, Can. J. Phys. 33: 693 (1955).
6. A. T. Blades and H. G. Thode, Naturforsch. 10A: 838 (1955).
7. A. T. Blades, W. H. Fleming, and H. G. Thode, Can. J. Chem. 34: 233 (1956).
8. J. E. Sattizahn, M. Kahn, and J. D. Knight, Bull. Am. Phys. Soc., Ser. 2, 2: 197 (1957).
9. A. F. Stehney and N. Sugarman, Phys. Rev. 89: 194 (1953).
10. G. W. Reed and A. Turkevich, Phys. Rev. 92: 1473 (1953).
11. A. P. Baerg and R. M. Bartholomew, Can. J. Chem. 35: 980 (1957).
12. C. D. Coryell, A. Sakakura, and A. M. Ross, Phys. Rev. 77: 755 (1950).
13. W. H. Hardick, Phys. Rev. 92. 1072 (1953).
14. B. C. Purkayastha and G. R. Martin, Can. J. Chem. 34: 293 (1956).
15. A. C. Pappas and D. R. Wiles, J. Inorg. Nucl. Chem. 2: 69 (1956).
16. R. M. Bartholomew, et al., Can. J. Chem. 31: 120 (1953).
17. S. Katcoff and W. Rubinson, Phys. Rev. 91: 1458 (1953).
18. L. Yaffe, A. E. Day, and B. A. Greer, Can. J. Chem. 31: 48 (1953).
19. A. C. Wahl, Phys. Rev. 99: 730 (1955).
20. F. Brown, J. Inorg. Nucl. Chem. 1: 248 (1955).
21. R. M. Bartholomew and A. P. Baerg, Can. J. Chem. 34: 201 (1956).
22. L. Yaffe, et al., Can. J. Chem. 32: 1017 (1954).
23. H. G. Petrow and G. Rocco, Phys. Rev. 96: 1614 (1954).
24. E. C. Freiling, L. R. Bunney, and N. E. Ballou, Phys. Rev. 96: 102 (1954).

---

* No references to these items appear in the text.

25. A. Pappas, Geneva 1955 conference.

26. T. J. Kennet and H. G. Thode, Phys. Rev. 103: 323 (1956).

27. K. F. Flynn, L. E. Glendenin, and E. P. Steinberg, Phys. Rev. 101: 1942 (1956).

28. J. E. Brolley, et al., Phys. Rev. 83: 990 (1951).

29. E. A. Melaika, et al., Can. J. Chem. 33: 830 (1955).

30. D. M. Wiles, J. A. Petruska, and R. H. Tomlinson, Can. J. Chem. 34: 227 (1956).

31. L. M. Krizhanskii, et al., At. Energ. 2 (3): 276 (1957).

32. W. H. Fleming and H. G. Thode, Can. J. Chem. 34: 193 (1956).

33. R. K. Wanless and H. G. Thode, Can. J. Phys. 33: 541 (1955).

34. T. J. Kennet and H. G. Thode, Can. J. Phys. 35: 969 (1957).

35. J. E. Simmons, R. L. Henkel, and J. E. Brolley, Bull. Am. Phys. Soc., Ser. 2, 2: 308 (1957).

36. P. K. Kuroda and R. R. Edwards, J. Chem. Phys. 22: 1940 (1954).

37. L. E. Glendenin and C. D. Coryell, Phys. Rev. 77: 755 (1950).

38. A. B. Smith, A. M. Friedman, and P. R. Fields, Phys. Rev. 102: 813 (1956).

39. W. E. Grummit and G. Wilkinson, Nature 161: 520 (1948).

40. W. E. Grummitt, J. Gueron, G. Wilkinson, and L. Yaffe, Can. J. Research 25B: 364 (1947).

41. M. G. Inghram, D. C. Hess, and J. H. Reynolds, Phys. Rev. 76: 1717 (1949).

42. H. G. Thode and R. B. Shields, Rept. Progr. Phys. 12: 18 (1949).

43. L. Yaffe and C. E. Mackintosh, Can. J. Research 25B: 371 (1947).

44. E. W. Titterton and T. A. Brinkley, Phil. Mag. 41: 500 (1950).

45. U. R. Arifkhanov, M. M. Makarov, N. A. Perfilov, and V. P. Shamov, Zh. Eksperim. i Teor. Fiz. 38: 1115 (1960).

46. A. Kjelberg and A. C. Pappas, J. Inorg. Nucl. Chem. 11: 173 (1959).

47. F. T. Ashizawa and P. K. Kuroda, J. Inorg. Nucl. Chem. 5: 12(1957).

48. L. M. Krizhanskii and A. N. Murin, At. Energ. 4 (1): 77 (1958).

49. C. B. Fulmer and B. L. Cohen, Phys. Rev. 108: 370 (1957).

50. D. L. Hill, Proc. of the Second United Nations Inter. Conf. on the Peaceful Uses of Atomic Energy 15: 244, P/660, Geneva, 1958.

51. W. J. Swiatecki, Proc. of the Second United Nations Inter. Conf. on the Peaceful Uses of Atomic Energy 15: 248, P/651, Geneva, 1958.

52. A. Hemmindinger, Geneva 1958 conference.

53. A. Smith, et al., Proc. of the Second United Nations Inter. Conf. on the Peaceful Uses of Atomic Energy 15: 392, P/690, Geneva, 1958.

54. A. G. W. Cameron, Proc. of the Second United Nations Inter. Conf. on the Peaceful Uses of Atomic Energy 15: 425, P/198, Geneva, 1958.

55. B. T. Geilikman, Geneva 1958 conference.

56. B. T. Geilikman, Geneva 1958 conference.

57. V. G. Nosov, Geneva 1955 conference.

58. D. Strominger, J. M. Hollander, and G. T. Seaborg, Rev. Mod. Phys. 30: 585 (1958).

59. V. K. Gorshkov and M. P. Anikina, At. Energ. 7 (2): 144 (1959).

60. M. G. Mayer, Phys. Rev. 74: 235 (1948).

61. L. Meitner, Arkiv. Fysik 4: 383 (1952).

62. R. D. Hill, Phys. Rev. 98: 1272 (1955).

63. V. Vladimirskii, Zh. Eksperim. i Teor. Fiz. 32: 822 (1957).

64. O. Hahn and F. Srassmann, Naturwissenschaften 27: 11 (1939).

65. O. Hahn and F. Srassmann, Naturwissenschaften 27: 529 (1939).

66. F. A. Heyn, A. H. Aten, and C. J. Bakker, Nature 143: 516 (1939).

67. R. W. Dodson and R. D. Fowler, Phys. Rev. 55: 880 (1939).

68. P. Abelson, Phys. Rev. 56; 1 (1939).

69. A. Langsdorf, Phys. Rev. 56: 205 (1939).

70. L. Meitner and O. R. Frisch, Nature 143: 239 (1939).

71. O. R. Frisch, Nature 143: 276 (1939).

72. G. N. Glasoe and J. Steigman, Phys. Rev. 58: 1 (1940).

73. Y. Nishina, et al., Nature 146: 24 (1940).

74. E. Segre and C. S. Wu, Phys. Rev. 57: 552 (1940).

75. Y. Nishina, et al., Phys. Rev. 58; 660 (1940).

76. R. W. Dodson and R. D. Fowler, Phys. Rev. 57; 966 (1940).

77. V. G. Khlopin, Izv. Akad. Nauk SSSR, Ser. Fiz. 4 (2): 305 (1940).

78. H. L. Anderson, E. Fermi, and A. V. Grosse, Phys. Rev. 59: 52 (1941).

79. Y. Nishina, et al., Phys. Rev. 59: 323 (1941).

80. Y. Nishina, et al., Z. Phys. 119: 195 (1942).

81. W. Jentschke and F. Prankl, Z. Phys. 119: 696 (1942).

82. W. Jentschke, Z. Physik 120: 165 (1943).

83. A. Moussa and L. Goldstein, Compt. rend. acad. sci., Paris 212: 986 (1941).

84. D. V. Hanna, Seminar "Structure of the Nucleus," Moscow, Atom Press, 1962, p. 322.

85. A. H. W. Aten, Jr., Physica 28 (3): 262 (1962).

86. P. Fong, Phys. Rev. 122 (5): 1543 (1961).

87. J. Tarrell, Phys. Rev. 127 (3): 880 (1962).

88. H. W. Newson, Phys. Rev. 122 (4): 1224 (1961).